# 內容提要

　　AI 人工智慧繪畫工具的使用難點在於關鍵字（prompt）的描述。如何寫出更為精准的關鍵字描述，從而讓畫面達到自己想要的效果才是重中之重。

　　本書針對對 AI 技術好奇的讀者，以 Midjourney 軟體為主，整理了一些當今流行的 AI 繪畫關鍵字，按照精度渲染、構圖視角、風格類型、色彩質感、光照效果、場景呈現等主題分類，每個分類下都詳細講解了此類關鍵字的詞語名稱、圖片效果、提示詞、重點詞語等，以幫助大家更快地選取恰當、準確的 AI 繪畫關鍵字。

　　如果你能掌握這些關鍵字，並且靈活地應用到 AI 繪畫中，那麼你就可以「妙語生畫」了！

# 使用說明

❶ 詞語名稱
每個詞語的中文和英文名稱

❷ 圖片效果
輸入該詞語會出現的相關圖片效果

## 渲染　Maxon cinema 4D

prompt: Maxon cinema 4D,colorful character a vibrant environment, dynamic scenes and camera movements, neon style lighting effects, sci-fi and fantasy elements --ar 16:9 --v 5.1

## 建築渲染　Architectural visualisation

pro Architectural visualisation, clean and simple lines with a focus on functionality and efficiency, natural materials such as wood or stone, in a modern and sleek style --ar 16:9 --v 5.1

❸ 提示詞
達到圖片效果所需
的完整關鍵字

❹ 重點詞語
提示詞中的重點詞語用彩色
標記出來,一目了然

❺ 頁碼
每一頁面上標明次序的
數字,方便檢索

# 目錄 ━━━━

## 第 1 章　精度渲染　　　　　　　　　　PAGE 009

高動態 .................... 010
超高清 .................... 010
全高清 .................... 011
1080P 解析度 ............ 011
2K 解析度 ................ 012
4K 解析度 ................ 012
8K 解析度 ................ 013
高細節 .................... 013
超高品質 ................. 014

高解析度 ................. 014
最佳品質 ................. 015
傑作 ....................... 015
聚焦清晰 ................. 016
更多詳細的細節 ......... 016
超精細繪畫 .............. 017
極其詳細的刻畫 ......... 017
虛幻引擎 ................. 018
OC 渲染 .................. 018

渲染 ....................... 019
建築渲染 ................. 019
室內渲染 ................. 020
真實感 .................... 020
V 射線 .................... 021
基於物理渲染 ........... 021
渲染效果 ................. 022
虛擬引擎 5 ............... 022

## 第 2 章　構圖視角　　　　　　　　　　PAGE 023

飽和構圖 ................. 024
視角構圖 ................. 024
剪影構圖 ................. 025
重複構圖 ................. 025
焦點構圖 ................. 026
對比構圖 ................. 026
重疊構圖 ................. 027
孤立構圖 ................. 027
輻射狀構圖 .............. 028
分割互補構圖 ........... 028
場面調度構圖 ........... 029
負空間構圖 .............. 029
拼貼構圖 ................. 030
對稱構圖 ................. 030
非對稱構圖 .............. 031
三分法構圖 .............. 031
黃金分割構圖 ........... 032
對角線構圖 .............. 032
動態對稱構圖 ........... 033
匯聚線條構圖 ........... 033
並列構圖 ................. 034
消失點構圖 .............. 034

非線性構圖 .............. 035
中心構圖 ................. 035
S 形構圖 ................. 036
橫向構圖 ................. 036
引導線 .................... 037
圓形構圖 ................. 037
鳥瞰圖 .................... 038
頂視圖 .................... 038
前視圖 .................... 039
後視圖 .................... 039
仰視圖 .................... 040
底視圖 .................... 040
第一人稱視角 ........... 041
微觀 ....................... 041
高角度視圖 .............. 042
核桃的橫截面圖 ......... 042
超側角 .................... 043
兩點透視 ................. 043
三點透視 ................. 044
立面透視 ................. 044
大特寫 .................... 045
胸部以上 ................. 045

頭部特寫 ................. 046
臉部特寫 ................. 046
膝蓋以上 ................. 047
全身 ....................... 047
半身像 .................... 048
細節鏡頭 ................. 048
中特寫 .................... 049
人在遠方 ................. 049
肖像 ....................... 050
頭部拍攝 ................. 050
特寫 ....................... 051
極端特寫 ................. 051
中景 ....................... 052
中遠景 .................... 052
遠景 ....................... 053
過肩鏡頭 ................. 053
鬆散景 .................... 054
近距離景 ................. 054
兩景 ....................... 055
三景 ....................... 055
風景拍攝 ................. 056
散景 ....................... 056

# CONTENTS

前景 ..................................... 057
背景 ..................................... 057
全長鏡頭 .............................. 058
全景視角 .............................. 058
超廣角 .................................. 059
等軸測圖 .............................. 059
微距 ..................................... 060

廣闊的視野 .......................... 060
逆光攝影風格 ...................... 061
寬景 ..................................... 061
超廣角鏡頭 .......................... 062
移軸 ..................................... 062
衛星視圖 .............................. 063
產品視圖 .............................. 063

極端特寫視圖 ...................... 064
電影鏡頭 .............................. 064
焦點對準 .............................. 065
景深 ..................................... 065
廣角鏡頭 .............................. 066
索尼阿爾法相機 .................. 066

## 第 3 章  風格類型                    PAGE 067

蒙太奇 .................................. 068
黑白 ..................................... 068
時尚 ..................................... 069
現代風格 .............................. 069
室內設計 .............................. 070
環保清新風格 ...................... 070
沉浸式設計 .......................... 071
鄉村風格 .............................. 071
現代簡約風格 ...................... 072
現代奢華風格 ...................... 072
草原風格 .............................. 073
森林風格 .............................. 073
海洋風格 .............................. 074
藝術裝飾風格 ...................... 074
珍珠奶茶風格 ...................... 075
中式風格 .............................. 075
新中式風格 .......................... 076
傳統中國水墨風格 .............. 076
東方山水畫 .......................... 077
水墨渲染 .............................. 077
水墨插圖 .............................. 078
水墨風格 .............................. 078
彩墨紙本 .............................. 079
紙本彩色顏料 ...................... 079
民族風格 .............................. 080
民族藝術 .............................. 080
傳統文化 .............................. 081
浮世繪 .................................. 081

日本漫畫風格 ...................... 082
日本動畫片 .......................... 082
像素風 .................................. 083
日本風格 .............................. 083
日本海報風格 ...................... 084
宮崎駿風格 .......................... 084
新海誠 .................................. 085
副島成記 .............................. 085
山田章博 .............................. 086
吉卜力風格 .......................... 086
JoJo 的奇妙冒險 ................. 087
ACGN 文化 .......................... 087
卡通 ..................................... 088
魔幻現實 .............................. 088
迪士尼風格 .......................... 089
蘿莉塔風格 .......................... 089
童話風格 .............................. 090
插畫 ..................................... 090
兒童插畫 .............................. 091
童話故事書插圖風格 .......... 091
水彩兒童插畫 ...................... 092
六七質 .................................. 092
向量圖 .................................. 093
油畫 ..................................... 093
攝影 ..................................... 094
水彩 ..................................... 094
素描 ..................................... 095
水墨畫 .................................. 095

雕塑 ..................................... 096
印刷版畫 .............................. 096
利諾剪裁 .............................. 097
手繪風格 .............................. 097
塗鴉 ..................................... 098
手稿 ..................................... 098
夢工廠動畫風格 .................. 099
夢工廠影業 .......................... 099
皮克斯 .................................. 100
賽博龐克 .............................. 100
黑色低俗 .............................. 101
好萊塢風格 .......................... 101
蒙太奇 .................................. 102
底片攝影風格 ...................... 102
微縮模型電影風格 .............. 103
巴洛克藝術 .......................... 103
法國藝術 .............................. 104
包浩斯 .................................. 104
達文西 .................................. 105
梵谷 ..................................... 105
克洛德・莫內 ...................... 106
新藝術風格 .......................... 106
洛可可 .................................. 107
文藝復興 .............................. 107
立體主義 .............................. 108
印象派 .................................. 108
點描畫派 .............................. 109
野獸派 .................................. 109

彩色玻璃窗.................110
抽象風.................110
孔版印刷風.................111
設計風.................111
歐普藝術.................112
光效應藝術.................112
次表面散射.................113
蒸汽龐克.................113
普普藝術.................114
寫實主義.................114
抽象表現主義.................115
超寫實主義.................115
超現實主義.................116
極簡主義.................116
歌德式.................117
歌德式黑暗.................117
粗獷主義.................118
建構主義.................118

未來主義.................119
工業風格.................119
書頁.................120
雕刻藝術風格.................120
美式鄉村風格.................121
真實的.................121
3D 風格.................122
複雜的.................122
真實感.................123
國家地理.................123
電影般的.................124
建築素描.................124
對稱肖像.................125
清晰的面部特徵.................125
局部解剖.................126
絎縫藝術.................126
復古黑暗.................127
90 年代電玩遊戲.................127

古典風格.................128
維多利亞時代.................128
遊戲風格.................129
神秘幻想.................129
魔幻現實主義風格.................130
角色概念藝術.................130
遊戲場景圖.................131
武器設計.................131
曠野之息.................132
戰甲神兵.................132
寶可夢.................133
Apex 英雄.................133
上古卷軸.................134
魂系遊戲.................134
底特律：變人.................135
劍與遠征.................135
跑跑薑餅人.................136
英雄聯盟.................136

# 第 4 章　色彩質感

紅.................138
橙.................138
黃.................139
綠.................139
青.................140
藍.................140
紫.................141
灰.................141
棕.................142
白.................142
黑.................143
薄荷綠色系.................143
日暮色系.................144
楓葉紅色系.................144
雪山藍色系.................145
雷射糖果紙色.................145
馬卡龍色.................146
莫蘭迪色系.................146
鈦金屬色系.................147
鮮果色系.................147
黑白灰色系.................148
極簡黑白色系.................148
溫暖棕色系.................149
柔和粉色系.................149
水晶藍色系.................150
發光.................150

星閃.................151
螢光.................151
聖光.................152
反射透明彩虹色.................152
糖果色系.................153
珊瑚色系.................153
薰衣草色系.................154
綠寶石色系.................154
玫瑰金色系.................155
淺藍色系.................155
酒紅色系.................156
藍綠色系.................156
黑色背景居中.................157
白色和綠色調.................157
紅色和黑色調.................158
黃色和黑色調.................158
金色和銀色調.................159
霓虹色調.................159
亮麗橙色系.................160
象牙白色系.................160
自然綠色系.................161
奢華金色系.................161
穩重藍色系.................162
經典紅黑白色系.................162
珊瑚橙色系.................163
巨無霸色系.................163

秋日棕色系.................164
丹寧藍色系.................164
時尚灰色系.................165
芭比粉色系.................165
紫羅蘭紫色系.................166
彩虹色系.................166
啞光質感.................167
珍珠質感.................167
綢緞質感.................168
毛絨質感.................168
水波紋質感.................169
珠光質感.................169
竹子質感.................170
金屬質感.................170
石頭質感.................171
石墨質感.................171
玻璃質感.................172
皮革質感.................172
塑膠質感.................173
水晶質感.................173
棉質.................174
沙質.................174
陶瓷質感.................175
紗綢質感.................175
磚石質感.................176
古銅質感.................176

油漆質感 ....................... 177
金屬漆質感 ................... 177
菌絲 ............................... 178
木頭 ............................... 178
腐朽衰敗的 ................... 179
骨骼狀 ........................... 179
玻璃 ............................... 180
棉花 ............................... 180
亞麻布 ........................... 181
雷絲 ............................... 181
瓷器 ............................... 182
青瓷 ............................... 182
琺瑯 ............................... 183
黏土質感 ....................... 183
紋理質感 ....................... 184
皮毛質感 ....................... 184
雕刻質感 ....................... 185
砂岩 ............................... 185
天鵝絨 ........................... 186
薄紙巾 ........................... 186

曲線細膩的 ................... 187
有層次感的 ................... 187
線條優美且精細的 ....... 188
線條簡潔且清晰的 ....... 188
有紋理的 ....................... 189
線條彎曲但流暢的 ....... 189
紋路自然的 ................... 190
大膽的顏色 ................... 190
具有浮雕感的 ............... 191
具有雕刻感的 ............... 191
超細節 ........................... 192
光滑的 ........................... 192
清晰的 ........................... 193
細膩的 ........................... 193
精細的 ........................... 194
平整的 ........................... 194
精密的 ........................... 195
線條流暢的 ................... 195
流線型的 ....................... 196
線條優美的 ................... 196

彎曲的 ........................... 197
柔和的 ........................... 197
多樣化的 ....................... 198
有機的 ........................... 198
光潔的 ........................... 199
微妙的 ........................... 199
纖細的 ........................... 200
細長的 ........................... 200
線條細緻的 ................... 201
線條流暢且柔和的 ....... 201
優雅的 ........................... 202
曲線優美的 ................... 202
療癒的 ........................... 203
猛烈的 ........................... 203
不規則的 ....................... 204
龐大的 ........................... 204
銳利的 ........................... 205
多稜角的 ....................... 205
充滿動感的 ................... 206
統一的 ........................... 206

# 第 5 章　光照效果

電影光 ........................... 208
戲劇燈光 ....................... 208
強光逆光 ....................... 209
立體光 ........................... 209
閃光燈光 ....................... 210
影棚光 ........................... 210
雙性照明 ....................... 211
反射 ............................... 211
柔和的照明 ................... 212
柔光 ............................... 212
投影 ............................... 213
發光 ............................... 213
螢光燈 ........................... 214
浪漫燭光 ....................... 214
柔和燭光 ....................... 215
好看的燈光 ................... 215
電光閃爍 ....................... 216
柔軟的光線 ................... 216
曖昧光暈 ....................... 217
自然光 ........................... 217
魔法森林 ....................... 218
霧氣朦朧 ....................... 218
夢幻霧氣 ....................... 219
仙氣繚繞 ....................... 219

溫暖光輝 ....................... 220
憂鬱氛圍 ....................... 220
柔和月光 ....................... 221
體積光 ........................... 221
逆光 ............................... 222
硬光 ............................... 222
林布蘭光 ....................... 223
輪廓光 ........................... 223
情調光 ........................... 224
晨光 ............................... 224
太陽光 ........................... 225
黃金時刻光 ................... 225
暗黑的 ........................... 226
鮮豔的 ........................... 226
暖光 ............................... 227
彩色光 ........................... 227
賽博龐克光 ................... 228
反光 ............................... 228
映射光 ........................... 229
氣氛照明 ....................... 229
殘酷的 ........................... 230
強烈對比的 ................... 230
陰影效果 ....................... 231
微光 ............................... 231

強光 ............................... 232
冷光 ............................... 232
明亮的 ........................... 233
雲隙光 ........................... 233
外太空觀 ....................... 234
分割布光 ....................... 234
前燈 ............................... 235
背光照明 ....................... 235
側光 ............................... 236
邊緣光 ........................... 236
頂光 ............................... 237
乾淨的背景趨勢 ........... 237
邊緣燈 ........................... 238
全域照明 ....................... 238
霓虹燈冷光 ................... 239
明暗分明 ....................... 239
黑暗氛圍 ....................... 240
鮮豔色彩 ....................... 240
高對比度 ....................... 241
安靜恬淡 ....................... 241
明亮高光 ....................... 242
閃耀星空 ....................... 242

# 第 6 章 場景呈現

| | | |
|---|---|---|
| 反烏托邦 244 | 巨大建築 265 | 神秘寺廟 286 |
| 幻想 244 | 賽博龐克小巷 265 | 古代遺跡 286 |
| 異想天開 245 | 星空夜景 266 | 沙漠綠洲 287 |
| 廢墟 245 | 數位宇宙 266 | 月球殖民地 287 |
| 教室 246 | 超現實夢境 267 | 蒸汽動力機械 288 |
| 臥室 246 | 太空船 267 | 廢棄太空船 288 |
| 森林 247 | 懸崖峭壁 268 | 神秘古墓 289 |
| 城市 247 | 神秘森林 268 | 冥界 289 |
| 足球場 248 | 天空島嶼 269 | 霓虹城市 290 |
| 體育場 248 | 水晶宮殿 269 | 外星球 290 |
| 競技場 249 | 荒漠孤煙 270 | 未來公園 291 |
| 摔跤場 249 | 沉船遺跡 270 | 時光穿越城市 291 |
| 泳池 250 | 仙人掌沙漠 271 | 機器人工廠 292 |
| 咖啡廳 250 | 神話世界 271 | 賽博龐克叢林 292 |
| 麵包店 251 | 外太空 272 | 神秘山脈 293 |
| 書店 251 | 魔幻森林 272 | 冰洞穴 293 |
| 居酒屋 252 | 古代神廟 273 | 賽博龐克城市 294 |
| 宴會 252 | 火山噴發 273 | 反烏托邦未來 294 |
| 音樂會 253 | 未來機器人 274 | 鬼影森林 295 |
| 植物園 253 | 巨大機器 274 | 外星地貌 295 |
| 遊樂園 254 | 末日廢墟 275 | 超現實主義風景 296 |
| 教堂 254 | 星際大戰 275 | 中世紀城堡 296 |
| 中式閣樓 255 | 火星探險 276 | 月球地貌 297 |
| 花海 255 | 科技城市 276 | 水晶洞穴 297 |
| 廢棄城市建築群 256 | 浪漫小鎮 277 | 世界末日城市 298 |
| 近未來都市 256 | 蒸汽龐克工廠 277 | 水下世界 298 |
| 街景 257 | 雨天城市 278 | 未來都市 299 |
| 煉金室 257 | 蘑菇森林 278 | 魔法花園 299 |
| 宇宙 258 | 童話城堡 279 | 魔法城堡 300 |
| 雨天 258 | 迷人花園 279 | 夢幻雲彩 300 |
| 在晨霧中 259 | 後啟示錄世界 280 | 工業城市 301 |
| 充滿陽光 259 | 魔法王國 280 | 歌德式大教堂 301 |
| 銀河 260 | 反烏托邦景觀 281 | 浮空城市 302 |
| 地牢 260 | 黑暗森林 281 | 火星 302 |
| 星雲 261 | 月球景觀 282 | 幽靈莊園 303 |
| 瘋狂麥斯風格 261 | 迷失廢墟 282 | 海盜島嶼 303 |
| 巴比倫空中花園 262 | 冰雪王國 283 | 未來實驗室 304 |
| 草原草地 262 | 熱帶天堂 283 | 太空站 304 |
| 雜草叢生的 263 | 後啟示錄荒野 284 | |
| 後啟示錄、末日後 263 | 蒸汽龐克城市 284 | |
| 天空之城 264 | 奇幻村莊 285 | |
| 北極光 264 | 賽博龐克街道 285 | |

# 精度渲染

第 1 章

精度渲染是電腦圖學中的一個術語,用於描述圖形或圖像渲染的精細程度。它涉及繪製和顯示圖形時所使用的資料的精確程度和準確性。渲染引擎通常提供了各種精度設置的選項,以便使用者根據具體需求進行調整。例如電子遊戲通常會選擇較低的精度以獲得更高性能,而在高品質的渲染任務中,例如電影特效製作,通常會追求更高的精度以獲得更逼真的效果。

## 高動態　HDR

prompt: HDR, bright, vibrant colors, surrealistic landscape, dreamy atmosphere, pastel hues --ar 16:9 --v 5.1

## 超高清　UHD

prompt: UHD, nature scene, vastness, depth of field, aerial view, high resolution, vibrant colors --ar 16:9 --v 5.1

## 全高清 FHD

prompt: FHD, A dreamy, hazy landscape of a river valley surrounded by mountains, in the style of romanticism, with an emphasis on the clouds and misty atmosphere, captured with a medium format film camera --ar 16:9 --v 5.1

## 1080P 解析度 1080P

prompt: 1080P, A young girl in a pink dress, standing on a beach, with the sun setting behind her, in the style of romanticism, warm colors, soft focus --ar 16:9 --v 5.1

## 2K 解析度　2K

prompt: 2K, A tranquil lake surrounded by mountains, golden hour light, soft focus, shallow depth of field, warm colors, impressionist style, high contrast --ar 16:9 --v 5.1

## 4K 解析度　4K

prompt: 4K, A young girl in a format, surrounded by a vibrant and colorful forest, with a dreamy look on her face. Hazy sunlight filters through the trees and casts shadows across the forest floor. --ar 16:9 --c 40 --v 5.1

## 8K 解析度　8K

prompt: 8K, A white kitten, in a dreamy atmosphere, with a shallow depth of field, in the style of surrealism, bright and vibrant colors --ar 16:9 --c 50 --v 5.1

 ## 高細節　High detail

prompt: High detail, 16:9, close-up of a flower in full bloom, pastel colors, light and airy, classical painting style, Rembrandt, soft focus --ar 16:9 --v 5.1

## 超高品質　Hyper quality

prompt: Hyper quality, 16:9, a person in a futuristic cityscape, with high tech buildings and neon lights, in the style of cyberpunk, sci-fi noir, --ar 16:9 --v 5.1

## 高解析度　High resolution

prompt: High resolution,16:9, detailed landscape with a lake and mountain in the background, dramatic sky, warm colors, aerial view, drone shot --ar 16:9 --v 5.1

## 最佳品質　Best quality

prompt: Best quality, close-up of a person's hands holding a cup of hot coffee while looking --ar 16:9 --v 5.1

## 傑作　Masterpiece

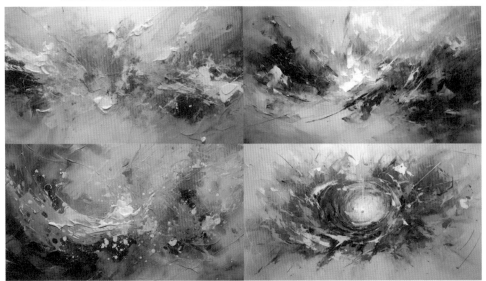

prompt: Masterpiece, painting, abstract, surrealistic, vibrant colors, brush strokes, depth, texture, gold and silver accents --ar 16:9 --v 5.1

## 聚焦清晰　Sharp focus

prompt: Sharp focus, woman wearing a red dress, standing in a field of wildflowers, sun setting in the background, vibrant colors, soft focus, shallow depth of field --ar 16:9 --v 5.1

## 更多詳細的細節　Highly detailed

prompt: Highly detailed, close-up shot of a leaf, in the style of macro photography, with blurred background, vibrant colors, and textures --ar 16:9 --v 5.1

## 超精細繪畫　Ultra-fine painting

prompt: Ultra-fine painting, a woman in a dress of golden yellow and pink, in the style of impressionism, a garden full of flowers, pastel colors, light and shadows, --ar 16:9 --v 5.1

## 極其詳細的刻畫　Extreme detail description

prompt: Extreme detail description, a closeup of a butterfly with intricate details and patterns, in the style of realism, macro lens, vibrant colors, muted background, --ar 16:9 --v 5.1

## 虚幻引擎　Unreal engine

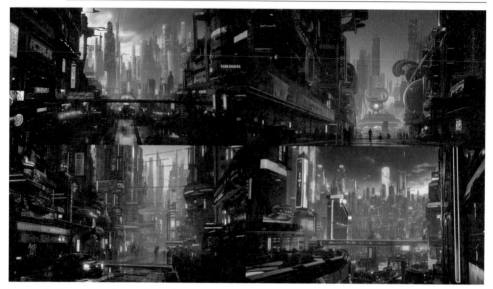

prompt: Unreal engine, with a futuristic city in the background, a high tech building in the foreground, neon lights, and a low angle perspective --ar 16:9 --v 5.1

## OC 渲染　Octane render

prompt: Octane render, a futuristic cityscape with robots, vibrant color palette, sharp details, deep shadows and highlights, 3D elements, sci-fi look --ar 16:9 --c 50 --v 5.1

## 渲染　Maxon cinema 4D

prompt: Maxon cinema 4D,colorful characters in a vibrant environment, dynamic scenes and camera movements, neon style lighting effects, sci-fi and fantasy elements --ar 16:9 --v 5.1

## 建築渲染　Architectural visualisation

prompt: Architectural visualisation, clean and simple lines with a focus on functionality and efficiency, natural materials such as wood or stone, in a modern and sleek style --ar 16:9 --v 5.1

## 室內渲染　Corona render

prompt: Corona render, warm and cozy atmosphere, living room, wooden furniture, comfortable sofa, small coffee table --ar 16:9 --v 5.1

## 真實感　Quixel megascans render

prompt: Quixel megascans render, a vast desert landscape, with mountains in the distance and sand dunes in the foreground, in a sepia tone, with a cinematic look --ar 16:9 --v 5.1

## V 射線　V-ray

prompt: V-ray, futuristic city, highly detailed, with a lot of vibrant colors, in the style of Blade Runner, --ar 16:9 --v 5.1

## 基於物理渲染　Physically-based rendering

prompt: Physically-based rendering, a realistic 3D environment with realistic lighting and shadows, high resolution textures and detailed geometry, --ar 16:9 --v 5.1

渲染效果　Vray rendering

prompt: Vray rendering, majestic dragon, shimmering scales, fiery breath, fierce eyes, soaring wingspan, mountainous landscape, ancient ruins in the background, mystical symbols etched on the ground, epic battle scene, realistic render style --ar 16:9 --v 5.1

 虛擬引擎 5　Unreal engine 5

prompt: Unreal engine 5, mystical forest, misty atmosphere, ancient ruins, overgrown vines and moss, hidden waterfalls and streams, magical creatures lurking in the shadows, ethereal and dreamlike quality, in a fantasy style --ar 16:9 --v 5.1

# 構圖視角

第2章

構圖視角是指攝影或繪畫作品中所採用的觀察角度或
視角。它決定了觀眾所看到的場景和主題的呈現方式，
對整體畫面的表現力和視覺效果有著重要影響。不同
的構圖視角可以帶來不同的觀感和情感體驗，我們可
以根據主題和表現意圖選擇合適的構圖視角。

## 飽和構圖　Saturated composition

prompt: Saturated composition, a red and yellow sunflower, in the style of surrealism, with vibrant colors, detailed petals, and a shallow depth of field --ar 16:9 --v 5.1

## 視角構圖　Point of view composition

prompt: Point of view composition, an abandoned city, dark and gloomy atmosphere, with a hint of mystery, streetlights in the distance, neon lights in the foreground, a sense of despair --ar 16:9 --v 5.1

## 剪影構圖　Cut out composition

prompt: Cut out composition, A person standing in silhouette against the warm-lit sky at sunset, shallow depth of field, soft focus, lens flare --ar 16:9 --v 5.1

 ## 重複構圖　Repetition composition

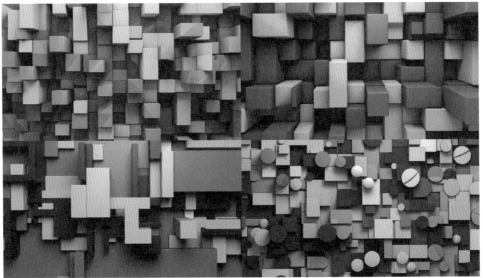

prompt: Repetition composition, abstract geometric shapes, minimalism style, with blue and yellow colors, very detailed --ar 16:9 --v 5.1

## 焦點構圖　Focal point composition

prompt: Focal point composition, a landscape with a mountain peak and a lake, light and shadows, pastel colors, high contrast, film camera, medium format --ar 16:9 --v 5.1

## 對比構圖　Contrast composition

prompt: Contrast composition, two elements in a frame, one in the foreground and the other in the background, with a strong contrast between them, minimalistic style, black and white color scheme, film camera --ar 16:9 --v 5.1

## 重疊構圖　Overlapping composition

prompt: Overlapping composition, monochromatic tones, geometric patterns, strong shadows and highlights, high contrast, industrial setting, grungy texture, vintage style, in a rectangular format --ar 16:9 --v 5.1

## 孤立構圖　Isolated composition

prompt: Isolated composition, stark contrast, monochromatic color scheme, sharp lines and angles, negative space, high-key lighting, in a modern art style, using Adobe Illustrator --ar 16:9 --v 5.1

## 輻射狀構圖　Radial composition

prompt: Radial composition, floral patterns, pastel colors, organic shapes, soft and delicate edges, romantic atmosphere, vintage texture, natural lighting, in a bohemian style --ar 16:9 --v 5.1

 ## 分割互補構圖　Split complementary composition

prompt: Split complementary composition, vibrant sunset, silhouetted palm trees, warm tones of orange and pink, contrasting cool tones of blue and purple in the sky, split complementary composition --ar 16:9 --v 5.1

**場面調度構圖　Mise-en-scene composition**

prompt: Mise-en-scene composition,vintage movie scene, dimly lit room, antique furniture, sepia-toned filter, a woman in a flowing dress sitting at a vanity table, a man in a suit, holding a cigarette in his hand, film noir style --ar 16:9 --v 5.1

 **負空間構圖　Negative space composition**

prompt: Negative space composition, monochromatic color scheme, minimalist style, sharp geometric shapes, high contrast, black and white, abstract concept of balance and harmony, in vector graphic style --ar 16:9 --v 5.1

## 拼貼構圖　Collage composition

prompt: Collage composition, vintage newspaper clippings, torn edges, overlapping layers, handwritten notes and doodles, faded colors, old photographs, mixed media, eclectic style --ar 16:9 --v 5.1

 ## 對稱構圖　Symmetrical composition

prompt: Symmetrical composition, minimalist design, monochromatic color scheme, sharp lines and angles, geometric shapes, high contrast lighting, futuristic atmosphere, digital art style, by Beeple --ar 16:9 --v 5.1

 非對稱構圖　Asymmetrical composition

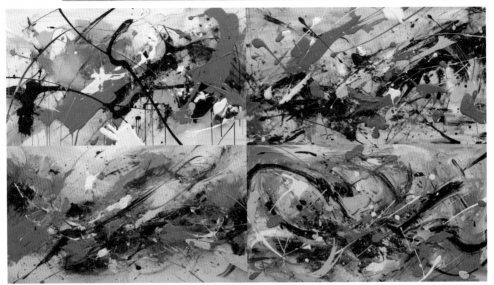

prompt: abstract art, Asymmetrical composition, vibrant colors, bold brushstrokes, organic shapes, dynamic movement, acrylic on canvas, by Jackson Pollock --ar 16:9 --v 5.1

## 三分法構圖　Rule of thirds composition

prompt: Rule of thirds composition, natural landscape, vibrant colors, golden hour lighting, mountain range in the background, a winding river in the foreground, a lone tree in the left third of the frame, birds flying in the sky --ar 16:9 --v 5.1

## 黃金分割構圖　Golden ratio composition

prompt: Golden ratio composition, serene mountain landscape, misty atmosphere, pastel color palette, soft and delicate brushstrokes, impressionist style, oil on canvas, by Claude Monet --ar 16:9 --v 5.1

## 對角線構圖　Diagonal composition

prompt: Diagonal composition, black and white photo, vintage feel, classic car, chrome accents, sleek lines, dramatic shadows and highlights, film noir atmosphere, wide angle lens, by Ansel Adams --ar 16:9 --v 5.1

## 動態對稱構圖　Dynamic symmetry composition

prompt: Dynamic symmetry composition, bold and contrasting colors, sharp lines and angles, geometric shapes, abstract forms, minimalist style, high contrast black and white --ar 16:9 --v 5.1

## 匯聚線條構圖　Converging lines composition

prompt: Converging lines composition, black and white, high contrast, diagonal perspective, urban architecture, minimalist style, stark and dramatic atmosphere, inky shadows and bright highlights --ar 16:9 --v 5.1

## 並列構圖　Juxtaposition composition

prompt: Juxtaposition composition, natural elements, organic shapes, warm tones, soft lighting, blurred edges, negative space, juxtaposed with industrial elements like metal pipes and concrete walls, in a botanical garden greenhouse --ar 16:9 --v 5.1

 ## 消失點構圖　Vanishing point composition

prompt: Vanishing point composition, natural landscape, serene atmosphere, winding road leading to a distant mountain range, soft pastel colors of the sky and clouds at sunset or sunrise, lush greenery and trees, birds flying in the distance --ar 16:9 --v 5.1

## 非線性構圖　Nonlinear composition

prompt: Nonlinear composition, vibrant colors, fluid shapes, organic textures, dynamic movement, energetic brushstrokes, acrylic paint on canvas, inspired by Wassily Kandinsky's style, in a modern and bold fashion --ar 16:9 --v 5.1

## 中心構圖　Center the composition

prompt: Center the composition, minimalist design, monochromatic color scheme, sleek and modern lines, high contrast lighting, symmetrical composition, in black and white photography style --ar 16:9 --v 5.1

## S 形構圖　S-shaped composition

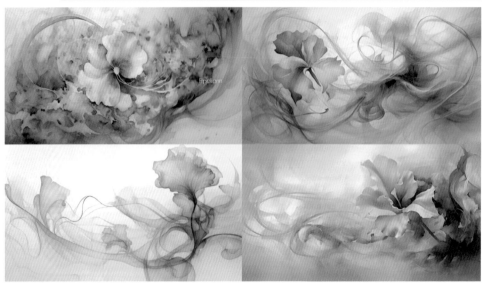

prompt: S-shaped composition, flowing lines, organic shapes, natural elements, soft pastel colors, dreamy atmosphere, watercolor style, asymmetrical balance, focal point on a flower in the foreground, by Claude Monet --ar 16:9 --v 5.1

## 橫向構圖　Horizontal composition

prompt: Horizontal composition, majestic lion, golden mane, standing on a rocky outcrop overlooking a vast savannah, dramatic clouds in the sky, powerful and regal stance --ar 16:9 --v 5.1

## 引導線　Leading lines

prompt: Leading lines, soft and blurred, meandering through a dreamy landscape, creating a sense of movement and flow, pastel colors and gentle textures, romantic composition, square format, low contrast, reminiscent of a watercolor painting --ar 16:9 --v 5.1

## 圓形構圖　Circular composition

prompt: Circular composition, natural elements, earthy tones, soft lighting, bokeh effect, vintage camera feel, rustic scene, wooden textures --ar 16:9 --v 5.1

## 鳥瞰圖　Bird view

prompt: Bird view, two girls walking side by side, one wearing a flowy dress and the other in ripped jeans and a leather jacket, surrounded by colorful autumn leaves --ar 1:2 --v 5.1

## 頂視圖　Top view

prompt: Top view, a girl full-body, realistic portrait, standing in a realistic background --ar 1:2 --v 5.1

## 前視圖　Front view

prompt: Front view, a full-body, a girl standing in a natural landscape, in the style of realism --ar 1:2 --v 5.1

## 後視圖　Back view

prompt: Back view, a full-body, a girl standing in a natural landscape, in the style of realism --ar 1:2 --v 5.1

## 仰視圖　Up view/Look up

prompt: Up view, towering skyscrapers, reaching towards the sky, glass and steel facade, reflecting the clouds above,  bustling city life below, birds soaring overhead --ar 1:2 --v 5.1

 ## 底視圖　Bottom view

prompt: Bottom view, majestic mountain range, snow-capped peaks, fluffy clouds in the sky, tranquil lake in the foreground, colorful autumn foliage on the trees, sense of grandeur and awe-inspiring beauty --ar 1:2 --v 5.1

第一人稱視角　First-person view

prompt: First-person view, surrounded by lush greenery, misty atmosphere, vibrant flowers, chirping birds, gentle breeze, wooden cabin in the distance, warm sunlight filtering through the leaves, in a painterly impressionist style --ar 1:2 --v 5.1

 微觀　Microscopic view

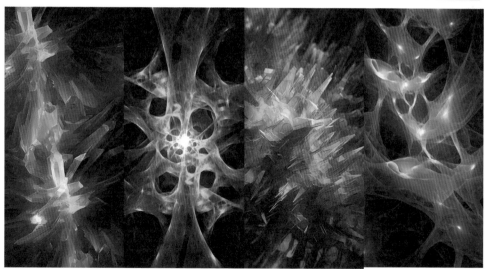

prompt: Microscopic view, crystal-like structures in shades of blue and purple, delicate filaments interweaving in a web-like formation, shimmering with iridescence, surrounded by a dark void, in digital art style, --ar 1:2 --v 5.1

高角度視圖　High angle view

prompt: High angle view, a bustling city, skyscrapers reaching towards the sky, busy streets filled with cars and people, green parks scattered throughout, sunlight casting long shadows, in a painterly style --ar 1:2 --v 5.1

 核桃的橫截面圖　A cross-section view of a walnut

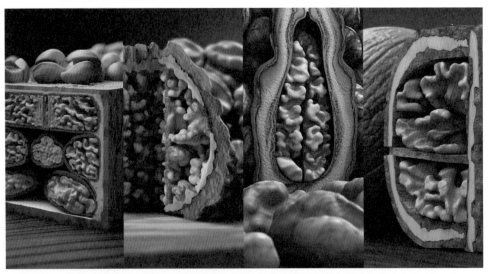

prompt: A cross-section view of a walnut, with detailed layers of shell and nutmeat, contrasting textures of smooth and rough, in a realistic hyper detailed render style, with a shallow depth of field to highlight the layers --ar 1:2 --v 5.1

## 超側角　Super side angle

prompt: Super side angle, punk rock musician, wild hair and makeup, leather jacket covered in studs and spikes, standing on stage in front of a roaring crowd, electric guitar in hand --ar 1:2 --v 5.1

## 兩點透視　Two-point perspective

prompt: Two-point perspective, geometric shapes, sharp edges, clean lines, minimalistic color palette, contrasting shadows, futuristic atmosphere, high contrast --ar 1:2 --v 5.1

## 三點透視　Three-point perspective

prompt: Three-point perspective, geometric shapes, monochromatic color scheme, sharp edges and clean lines, minimalist composition --ar 1:2 --v 5.1

## 立面透視　Elevation perspective

prompt: Elevation perspective, ancient temple ruins, overgrown with lush greenery, moss-covered stone steps leading up to the entrance --ar 1:2 --v 5.1

## 大特寫　Big close-up(BCU)

prompt: Big close-up, intense eyes, piercing gaze, dramatic lighting, sweat-drenched skin, furrowed brows, rugged stubble, veins popping out, raw emotion, in black and white film noir style --ar 1:2 --v 5.1

## 胸部以上　Chest shot

prompt: Chest shot, a superhero, rippling muscles, tight costume, intense expression, dramatic lighting, smoke in the background, dynamic composition, comic book style, bold colors, by Jim Lee --ar 1:2 --v 5.1

## 頭部特寫　Headshot

prompt: Headshot, mysterious woman, piercing gaze, flowing hair, dark lipstick, dramatic lighting, black and white, film noir style, smoke in the background, tilted angle --ar 1:2 --v 5.1

## 臉部特寫　Face shot

prompt: Face shot, intense gaze, dramatic lighting, black and white, high contrast, sharp features, fierce expression, minimalistic composition, film noir style --ar 1:2 --v 5.1

## 膝蓋以上　Knee shot

prompt: Knee shot, fitted and sleek cocktail dress, black and white color block design, standing on a rooftop overlooking the city skyline at night, confident and sultry expression, film noir style --ar 1:2 --v 5.1

 ## 全身　Full body

prompt: Full body, ethereal goddess, flowing white dress, billowing chiffon, bare feet, golden halo, surrounded by mist and flowers, soft pastel colors, dreamlike atmosphere --ar 1:2 --v 5.1

## 半身像　Busts

prompt: Busts, lifelike expressions and details, marble texture, dramatic lighting, classical Greek style, intricate hair and clothing details, in black and white photography style --ar 1:2 --v 5.1

## 細節鏡頭　Detail shot

prompt: Detail shot, extreme close-up, intricate textures and patterns, shallow depth of field, dramatic lighting, contrasting colors, minimalist composition, abstract and surreal atmosphere, in black and white film noir style --ar 1:2 --v 5.1

## 中特寫　MCU (Medium close-up)

prompt: MCU, a confident businessman, wearing a navy blue suit and a red tie, standing in front of a modern skyscraper, the sun shining brightly in the background, creating a lens flare effect, in a sleek and professional style --ar 1:2 --v 5.1

 ## 人在遠方　Extra long shot

prompt: Extra long shot, silhouettes against a fiery sunset, hazy and dreamy atmosphere, palm trees swaying in the wind, warm colors of orange, pink and yellow, composition with negative space, reminiscent of a postcard from a tropical paradise, by Steve McCurry --ar 1:2 --v 5.1

## 肖像　Portrait

prompt: Portrait, minimalist white background, subject with unique hairstyle and bold makeup, asymmetrical composition with subject off-center, bright neon lighting casting interesting shadows, playful expression and pose, digital art style with geometric shapes and patterns, by Shepard Fairey --ar 1:2 --v 5.1

## 頭部拍攝　Headshot

prompt: Headshot, piercing eyes, dramatic lighting, monochromatic color scheme, high contrast, bold shadows, minimalistic composition, stark background, in a film noir style --ar 1:2 --v 5.1

特寫　Close-up

prompt: Close-up, dewy skin, flushed cheeks, lips parted in anticipation, lashes fluttering, bokeh background, dreamy atmosphere, by Tim Walker --ar 1:2 --v 5.1

 極端特寫　Extreme close-up

prompt: Extreme close-up, intricate details, high contrast, dramatic lighting, black and white, intense emotion, sharp focus on eyes, minimalist composition, film noir style --ar 1:2 --v 5.1

## 中景　MS (Medium shot)

prompt: MS, adventurous traveler, backpack and camera gear, hiking in a mountain range, surrounded by lush greenery and wildflowers, breathtaking view of the valley below, warm sunlight filtering through the trees --ar 16:9 --v 5.1

 ## 中遠景　MLS (Medium long shot)

prompt: MLS, lonely traveler, standing on a cliff, overlooking a vast ocean, wind blowing through hair and clothes, rugged terrain, cloudy sky, distant horizon, vintage film style, warm color grading, grainy texture, 16mm camera, wide angle lens --ar 16:9 --v 5.1

## 遠景　LS (Long shot)

prompt: LS, vast desert landscape, sand dunes stretching to the horizon, a lone camel caravan traveling in the distance, clear blue sky with a few scattered clouds, warm and golden sunlight casting long shadows, peaceful and serene atmosphere --ar 16:9 --v 5.1

## 過肩鏡頭　Over the shoulder shot

prompt: Over the shoulder shot, serene expression, natural light, warm tones, bokeh effect, shallow depth of field, nature scene, mountains in the distance, Canon EOS R camera, 85mm lens, golden hour lighting, rule of thirds composition, soft focus --ar 16:9 --v 5.1

## 鬆散景　Loose shot

prompt: Loose shot, misty and dreamlike, vibrant flowers blooming in the underbrush, a gentle stream winding through the foliage, a lone deer grazing in a clearing, soft and diffused light filtering through the canopy --ar 16:9 --v 5.1

## 近距離景　Tight shot

prompt: Tight shot, fierce expression on a lion's face, sharp teeth bared, dramatic lighting casting deep shadows across the face, blurred background, wild and untamed atmosphere --ar 16:9 --v 5.1

## 兩景　2S (Two shot)

prompt: 2S, majestic mountain range, snow-capped peaks reaching towards the sky, a winding river flowing through the valley below, a lone hiker trekking through the wilderness, surrounded by breathtaking natural beauty and serenity --ar 16:9 --v 5.1

## 三景　3S (Three shot)

prompt: 3S, vintage steam locomotive train, intricate details and gears, steam billowing out of the chimney stack, traveling through a dense forest in autumn colors, wide angle view to capture the grandeur of the train --ar 16:9 --v 5.1

## 風景拍攝　Scenery shot

prompt: Scenery shot, breathtaking mountain view, crystal clear lake in the foreground, vibrant green trees and foliage, golden sunlight peeking through the clouds, high altitude perspective --ar 16:9 --v 5.1

## 散景　Bokeh

prompt: Bokeh, soft and gentle lights, pastel colors, heart-shaped shapes, floral elements in the background, romantic and whimsical style, composition centered, overall mood: sweet and charming --ar 16:9 --v 5.1

## 前景　Foreground

prompt: Foreground, elegant ballerina, graceful movements captured mid-air in a grand theater hall, captured in a close-up shot from below with shallow depth of field --ar 16:9 --v 5.1

## 背景　Background

prompt: Background, ethereal and dreamy, pastel colors blending together seamlessly, soft clouds floating in the sky, misty mountains in the distance, a tranquil lake reflecting the beauty of nature, birds chirping in the background, --ar 16:9 --v 5.1

## 全長鏡頭　Full length lens

prompt: Full length lens, urban street scene, rainy night, neon lights reflecting on the wet pavement, people rushing by with umbrellas and raincoats, taxi cabs passing by in a blur of color and light trails, in a gritty film noir style --ar 16:9 --v 5.1

 ## 全景視角　Panoramic view

prompt: Panoramic view, vast and majestic mountains, snow-capped peaks, crystal clear lakes, vibrant green forests, colorful wildflowers, golden sunset, warm and inviting atmosphere, peaceful and serene setting --ar 16:9 --v 5.1

## 超廣角　Utrawide shot

prompt: Utrawide shot, vast landscape, endless horizon, towering mountains in the distance, rolling hills and fields, clear blue sky, scattered clouds, birds flying in the distance, peaceful and serene atmosphere --ar 16:9 --v 5.1

## 等軸測圖　Isometric view

prompt: Isometric view, futuristic cityscape, neon lights, flying cars, towering skyscrapers, bustling streets, holographic advertisements, floating platforms, chrome and glass architecture, sleek and modern aesthetic, in a cyberpunk style --ar 16:9 --v 5.1

微距 Macro shot

prompt: Macro shot, crawling insect, intricate details, iridescent wings, creepy vibe, dark and moody background, sharp focus on the eyes and antennae, dramatic lighting from below, in black and white photography style --ar 16:9 --v 5.1

 廣闊的視野 An expansive view of

prompt: An expansive view of mountain range, vibrant colors of autumn foliage, distant lake reflecting the sky, clouds casting shadows over the landscape, hot air balloons dotting the sky --ar 16:9 --v 5.1

## 逆光攝影風格　Backlight style

prompt: Backlight style, silhouette subject, dramatic and contrasty lighting, warm
tones, lens flare, minimalist composition, negative space, ethereal atmosphere, inspired
by Gregory Crewdson --ar 16:9 --v 5.1

 ## 寬景　Wide view

prompt: Wide view, vast and endless horizon, rolling hills and valleys, dramatic cloud
formations, warm and vibrant color palette, painterly texture, impressionistic style --ar
16:9 --v 5.1

超廣角鏡頭　Ultra wide shot

prompt: Ultra wide shot, city skyline, towering skyscrapers, bustling streets, neon lights, traffic trails, reflective surfaces, modern and futuristic vibe, in cool and blue tones, with a hint of cyberpunk aesthetic --ar 16:9 --v 5.1

 移軸　Tilt-shift

prompt: Tilt-shift, miniature world, toy-like buildings and people, shallow depth of field, vibrant colors, dreamy atmosphere, soft focus, pastel tones, whimsical composition, fantasy style --ar 16:9 --v 5.1

## 衛星視圖　Satellite view

prompt: Satellite view, geometric patterns of crop fields, winding rivers and streams, mountain ranges in the distance, small villages and towns scattered throughout the landscape, soft pastel colors with a vintage feel --ar 16:9 --v 5.1

 ## 產品視圖　Product view

prompt: Product view, sleek and modern design, monochromatic color scheme, sharp edges and clean lines, soft lighting to highlight product features,  3D render in high resolution, realistic texture and materials, professional and polished look --ar 16:9 --v 5.1

極端特寫視圖　Extreme close-up view

prompt: Extreme close-up view of a diamond ring, sparkling facets and reflections, intricate metalwork details on the band, tiny engravings on the inside of the band, in high-end luxury product photography style, by Mario Testino --ar 16:9 --v 5.1

 電影鏡頭　Cinematic shot

prompt: Cinematic shot, dramatic lighting, intense shadows, high contrast, close-up of a person's face, piercing eyes, furrowed brow, sweat on the forehead, subtle smile, handheld camera movement, gritty and raw atmosphere --ar 16:9 --v 5.1

## 焦點對準　In focus

prompt: In focus, crystal clear details, macro shot, vibrant colors, natural light, shallow depth of field, bokeh effect, close-up of a flower petal, delicate veins and textures, dew drops glistening in the sun, soft and dreamy style --ar 16:9 --v 5.1

## 景深　Depth of field

prompt: Depth of field, selective focus, blurred foreground or background, prime lens with wide aperture for maximum bokeh effect or sharpness where needed, mixed lighting sources for contrast and interest --ar 16:9 --v 5.1

## 廣角鏡頭　Wide-angle view

prompt: Wide-angle view, vast landscape, sweeping view, endless horizon, rugged terrain, dramatic clouds, vibrant colors, wide-angle lens distortion, immersive experience --ar 16:9 --v 5.1

## 索尼阿爾法相機　Sony alpha camera

prompt: shot with a Sony alpha camera and 50mm lenselegant ballerina, wearing a white tutu and pointe shoes, dancing in a grand ballroom with a chandelier overhead, surrounded by other dancers in colorful costumes --ar 16:9 --v 5.1

# 風格類型

第**3**章

在藝術和設計領域，風格類型指的是不同的藝術風格
或設計風格。不同的風格類型通常具有獨有的特徵、
表現形式和審美標準，反映了不同的藝術觀點、文化
背景和歷史時期。例如文藝復興、抽象表現主義等，
每種風格類型都有其獨特的特徵和風格表達方式。藝
術家和設計師可以根據自己的創作意圖和審美偏好選
擇適合的風格類型來表達他們的想法。

蒙太奇　Montage

prompt: Montage, eclectic mix of images and textures, vintage photographs, hand-drawn illustrations, abstract shapes, bold colors, torn edges, playful and whimsical atmosphere, in a scrapbook style --ar 16:9 --v 5.1

 黑白　Black and white

prompt: Black and white, street photography, bustling city scene, candid moments, gritty textures, raw emotions, urban decay, crowded sidewalks and buildings --ar 16:9 --v 5.1

## 時尚 Fashion

prompt: Fashion, flowing silk dress, pastel colors, floral patterns, wide-brimmed hat, oversized sunglasses, vintage handbag, strappy heels, walking down a cobblestone street in Paris --ar 16:9 --v 5.1

## 現代風格 Modern-style

prompt: Modern-style, high contrast black and white color scheme, industrial materials and finishes, exposed brick walls and concrete floors, minimalist furniture with bold accents, edgy and urban vibe --ar 16:9 --v 5.1

**室內設計　Interior design**

prompt: Interior design, bold color scheme, mix of patterns and textures, vintage and modern elements, statement pieces of art and decor, unique lighting fixtures, unexpected details and accents, maximalist approach --ar 16:9 --v 5.1

 **環保清新風格　Eco-friendly-style**

prompt: Eco-friendly-style, natural fibers, earthy tones, loose and flowing silhouette, intricate embroidery details, handmade accessories, recycled materials, surrounded by lush greenery and blooming flowers, dreamy atmosphere --ar 16:9 --v 5.1

## 沉浸式設計　Immersive design

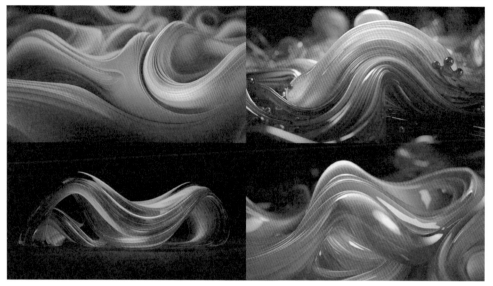

prompt: Immersive design, vibrant colors, flowing curves, organic shapes, interactive elements, dynamic lighting, futuristic technology, surreal atmosphere, 3D animation --ar 16:9 --v 5.1

## 鄉村風格　Country style

prompt: Country style girl, floral print dress, straw hat with ribbon, barefoot, holding a basket of freshly picked fruits and vegetables, standing in a lush green field, warm sunlight filtering through the trees --ar 16:9 --v 5.1

## 現代簡約風格　Modern minimalist

prompt: Modern minimalist, monochromatic color scheme, clean lines, simple shapes, natural materials, asymmetrical balance, soft lighting, abstract art on the walls, modern furniture --ar 16:9 --v 5.1

## 現代奢華風格　Modern luxury

prompt: Modern luxury, high-tech and innovative features, smart home automation system, virtual reality entertainment room, breathtaking views of the ocean and skyline, in a futuristic mansion --ar 16:9 --v 5.1

## 草原風格　Prairie

prompt: Prairie, golden wheat fields, scattered wildflowers, grazing horses and cattle, clear blue sky with fluffy white clouds, warm sunlight shining down, peaceful and serene atmosphere --ar 16:9 --v 5.1

## 森林風格　Forest

prompt: Forest, misty and mysterious, tall trees with twisted branches, vibrant wildflowers, sparkling streams, hidden creatures peeking out from behind tree trunks, soft sunlight filtering through the leaves --ar 16:9 --v 5.1

## 海洋風格　Ocean

prompt: Ocean, majestic whales swimming gracefully, schools of glittering fish darting around them, sunbeams shining through the water's surface creating a dreamy effect, tranquil atmosphere with a touch of mystery --ar 16:9 --v 5.1

## 藝術裝飾風格　Art deco style

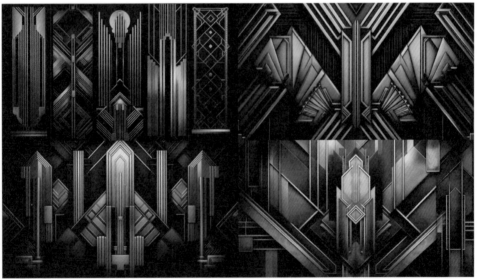

prompt: Art deco style, geometric shapes, metallic accents, bold colors, sleek lines, glamorous atmosphere, luxurious textures, symmetrical composition, high contrast, vintage feel --ar 16:9 --v 5.1

## 珍珠奶茶風格　Pearl milk tea style

prompt: Pearl milk tea style, frothy and bubbly, with a generous amount of pearls that overflow from the cup, served on a wooden tray with a small vase of fresh flowers on the side --ar 16:9 --v 5.1

## 中式風格　Chinese style

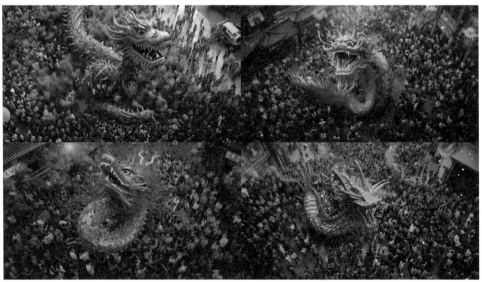

prompt: Chinese style, dragon dance performance, colorful dragon costume, intricate embroidery and beading, surrounded by a crowd of cheering people --ar 16:9 --v 5.1

## 新中式風格 New Chinese style

prompt: New Chinese style, garden, tranquil pond, colorful koi fish, blooming lotus flowers, winding pathways, ornate pavilions, intricate stone bridges, lush greenery --ar 16:9 --v 5.1

## 傳統中國水墨風格 Traditional Chinese ink painting style

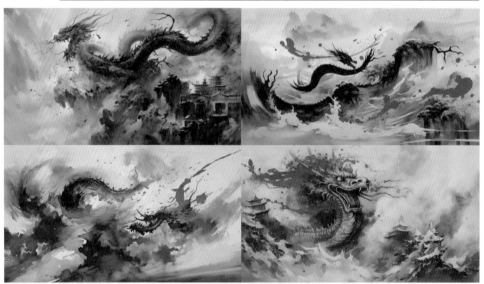

prompt: Traditional Chinese ink painting style, majestic dragon, swirling clouds, mountainous landscape, fierce expression, powerful wings, long tail --ar 16:9 --v 5.1

## 東方山水畫　Traditional Chinese ink painting

prompt: Traditional Chinese ink painting, misty mountains, flowing river, lone fisherman, bamboo forest, delicate brushstrokes, monochromatic color scheme --ar 16:9 --v 5.1

## 水墨渲染　Ink render

prompt: Ink render, flowing lines, abstract shapes, black and white contrast, organic and fluid composition, ethereal atmosphere --ar 16:9 --v 5.1

水墨插圖　Ink illustration

prompt: Ink illustration, intricate linework, monochromatic color scheme, twisted tree branches, hidden creatures lurking in the shadows, moonlit night sky --ar 16:9 --v 5.1

 水墨風格　Ink wash painting style

prompt: Ink wash painting style, serene mountain landscape, misty clouds, flowing river, lush greenery, towering trees, birds soaring in the sky, tranquil atmosphere, subtle brushstrokes --ar 16:9 --v 5.1

## 彩墨紙本　Color ink on paper

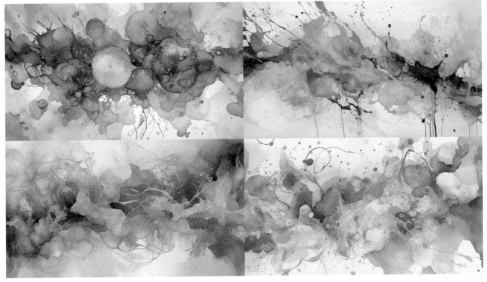

prompt: Color ink on paper, abstract shapes and lines, flowing and blending together, creating a vibrant and dynamic composition --ar 16:9 --v 5.1

## 紙本彩色顏料　Color pigment on paper

prompt: Color pigment on paper, abstract art, fluid and dynamic brushstrokes, bold and contrasting colors, intricately detailed patterns, overlapping layers --ar 16:9 --v 5.1

## 民族風格　Ethnic style

prompt: Ethnic style, vibrant and colorful headwrap, intricate braids, bold makeup, flowing dress with intricate patterns, standing in front of a lush green forest --ar 16:9 --v 5.1

## 民族藝術　Ethnic art

prompt: Ethnic art, bold and intricate patterns, warm and earthy tones, abstract shapes and forms, inspired by traditional tribal motifs, textured brushstrokes, in a contemporary art style --ar 16:9 --v 5.1

## 傳統文化　Traditional culture

prompt: Traditional culture, intricate wood carvings, delicate brushwork, subtle color palette, elegant calligraphy, serene garden setting --ar 16:9 --v 5.1

## 浮世繪　Japanese ukiyo-e

prompt: Japanese ukiyo-e, a woman in a kimono, surrounded by cherry blossoms and birds, with a bright sky and mountains in the background, --ar 16:9 --v 5.1

## 日本漫畫風格　Japanese comics style

prompt: Japanese comics style, a young girl in a kimono, standing in front of a bright cherry blossom tree, with a smile on her face, vibrant colors, delicate lines --ar 16:9 --niji 5

## 日本動畫片　Anime

prompt: Anime, a girl in a vibrant, colorful environment, with a surrealistic and dreamy atmosphere, in the style of Hayao Miyazaki, bright colors, soft lines, focus on the character's facial expression --ar 16:9 --niji 5

## 像素風　Pixel style

prompt: Pixel style, dynamic poses of characters and objects, stylized backgrounds and props, vivid colors and textures, strong contrast --ar 16:9 --niji 5

## 日本風格　Japanese style

prompt: Japanese style, cherry blossom petals floating in the air, traditional wooden house with paper sliding doors, tatami mat flooring, a tea set on a low table --ar 16:9 --niji 5

## 日本海報風格　Poster of Japanese graphic design

prompt: Poster of Japanese graphic design, bright and contrasting colors, playful typography, hand-drawn illustrations and patterns, layered composition --ar 16:9 --niji 5

## 宮崎駿風格　Hayao Miyazaki style

prompt: Hayao Miyazaki style, whimsical forest scene, vibrant colors, flying creatures with unique features, intricate details in every leaf and branch, soft and dreamy atmosphere, a hidden cottage in the middle of the forest --ar 16:9 --niji 5

## 新海誠　Makoto Shinkai

prompt: Makoto Shinkai, dreamy and surreal landscape, pastel colors, soft and gentle lighting, cherry blossom trees in bloom, distant mountains, a flowing river --ar 16:9 --niji 5

## 副島成記　Soejima Shigenori

prompt: Soejima Shigenori, pencil sketch, detailed and realistic facial features, flowing and dynamic hair, intense gaze, black and white color scheme, dramatic lighting, minimalist background, in the style of a classic portrait --ar 16:9 --niji 5

## 山田章博　Akihiro Yamada

prompt: Akihiro Yamada, traditional Japanese clothing, samurai armor, katana sword, cherry blossom background, stoic expression, dynamic pose, painted brushstroke style --ar 16:9 --niji 5

## 吉卜力風格　Ghibli studio

prompt: Ghibli studio, lush green forest, magical creatures, flying broomsticks, floating lanterns, cozy cottages, intricate details, warm and inviting atmosphere, hand-drawn animation style --ar 16:9 --niji 5

## JoJo 的奇妙冒險　JoJo's bizarre adventure

prompt: JoJo's bizarre adventure, muscular protagonist, flamboyant outfit, exaggerated poses and expressions, surreal environment with floating objects, metallic textures, dynamic composition with diagonal lines, high contrast, comic book style --ar 16:9 --niji 5

## ACGN 文化　ACGN(Animation、Comic、Game、Novel)

prompt: ACGN, vibrant hair color, oversized headphones, school uniform, holding a book or manga, surrounded by cherry blossom trees in full bloom, soft pastel color palette, dreamy and whimsical atmosphere --ar 16:9 --niji 5

## 卡通　Cartoon

prompt: Cartoon, round and chubby body, big eyes and smile, holding a bunch of colorful balloons, floating in a blue sky with fluffy clouds, happy and carefree atmosphere, in flat vector style --ar 16:9 --niji 5

## 魔幻現實　Magic realism

prompt: Magic realism, floating house with vibrant colors and intricate details, lush greenery surrounding it, a river flowing nearby, in dreamlike watercolor style --ar 16:9 --niji 5

## 迪士尼風格　Disney style

prompt: Disney style, fierce and adventurous warrior princess, bold colors, braided hair, feather headdress, mythical creature companion, ancient ruins setting, dramatic lighting, epic atmosphere, digital painting style --ar 16:9 --niji 5

## 蘿莉塔風格　Lolita style

prompt: Lolita style, pastel colors, delicate lace and ruffles, oversized bow, porcelain skin, big eyes, holding a parasol, surrounded by cherry blossom trees in full bloom, in a dreamy and romantic atmosphere --ar 16:9 --niji 5

## 童話風格　Fairy tale style

prompt: Fairy tale style, enchanted forest, magical creatures, glowing mushrooms, sparkling fairy dust, misty atmosphere, whimsical flowers, pastel colors, dreamlike composition, by Tim Walker --ar 16:9 --niji 5

## 插畫　Illustration

prompt: Illustration, fantastical creatures, vibrant colors, intricate details, dreamlike setting, floating islands, magical forests, flying birds and butterflies, watercolor style, by Hayao Miyazaki --ar 16:9 --niji 5

## 兒童插畫　Children's illustration

prompt: Children's illustration, adventurous young explorer, exploring a mysterious island filled with exotic creatures and hidden treasures, tropical jungle setting, backpack and binoculars in hand, accompanied by a loyal animal companion, --ar 16:9 --niji 5

## 童話故事書插圖風格　Stock illustration style

prompt: Stock illustration style, vibrant colors, fantastical creatures, magical landscapes, intricate details, hand-drawn style, enchanted forest setting, playful and dreamy atmosphere --ar 16:9 --v 5.1

## 水彩兒童插畫　Watercolor children's illustration

prompt: Watercolor children's illustration, soft pastel colors, cute animals and nature elements, happy and carefree children, using watercolor paper texture, with a touch of nostalgia and innocence --ar 16:9 --v 5.1

## 六七質　Munashichi

prompt: Munashichi, sleek black fur, piercing green eyes, sharp claws, standing on a rocky cliff overlooking the ocean, stormy clouds in the sky --ar 16:9 --v 5.1

## 向量圖　Vector

prompt: Vector, geometric shapes, intricate patterns, dynamic composition, futuristic atmosphere, neon lights, in a minimalist style with sharp lines and bold contrasts --ar 16:9 --v 5.1

## 油畫　Oil painting

prompt: Oil painting, floral still life, bold brushstrokes, rich colors, intricate details, dramatic lighting, classic composition, reminiscent of Van Gogh's style --ar 16:9 --v 5.1

## 攝影　Photography

prompt: Photography, black and white photo, dramatic lighting, portrait of an elderly person, wrinkles and lines on the face, deep set eyes, expressive expression, simple background, high contrast, raw emotion --ar 16:9 --v 5.1

## 水彩　Watercolor

prompt: Watercolor, bold and contrasting colors, expressive brushstrokes, capturing the essence of the subject's personality, playful and whimsical composition, with elements of nature and fantasy intertwined, evoking a sense of joy and wonder --ar 16:9 --v 5.1

素描　Sketch

prompt: Sketch, intricate patterns, fine details, soft shading, natural elements, organic shapes, harmonious composition, peaceful atmosphere, in pencil style --ar 16:9 --v 5.1

水墨畫　Ink painting

prompt: Ink painting, flowing water and mountains, misty atmosphere, delicate brushstrokes, monochromatic color scheme, minimalist composition, traditional Chinese style, by Wu Guanzhong --ar 16:9 --v 5.1

## 雕塑　Sculpture

prompt: Sculpture, graceful goddess, flowing robes, delicate features, serene expression, marble material, intricate carving details, outdoor garden setting, natural lighting, Neoclassical style --ar 16:9 --v 5.1

## 印刷版畫 · Block print

prompt: Block print, bright and bold hues, intricate and detailed designs, floral and botanical motifs, smooth and polished texture, romantic and whimsical feel, copperplate printing technique, centered composition, by William Morris --ar 16:9 --v 5.1

## 利諾剪裁　Lino cut

prompt: Lino cut, intricate patterns, overlapping shapes, muted colors, nature-inspired motifs, delicate linework, minimalist composition, negative space, modern art style, inspired by Matisse --ar 16:9 --v 5.1

## 手繪風格　Hand drawn style

prompt: Hand drawn style, vibrant colors, playful animals, hidden fairies, towering trees, dappled sunlight, watercolor style, loose brushstrokes, dreamlike atmosphere --ar 16:9 --v 5.1

## 塗鴉　Doodle

prompt: Doodle, colorful and whimsical, full of swirls and loops, featuring animals and nature elements, in a sketchbook style, with pencil strokes and smudges, creating a nostalgic feel, on a textured paper background --ar 16:9 --v 5.1

## 手稿　Manuscript

prompt: Manuscript, ancient parchment paper, faded ink, intricate calligraphy, illuminated letters, bound in leather with brass clasps, resting on a wooden lectern in a dimly lit library, surrounded by dusty tomes and flickering candles --ar 16:9 --v 5.1

## 夢工廠動畫風格　CGSociety style

prompt: CGSociety style, A young girl, standing in a surreal environment, with bright and vivid colors, --ar 16:9 --v 5.1

## 夢工廠影業　DreamWorks Pictures

prompt: DreamWorks Pictures, a bright and colorful landscape with a magical creature, in the style of a movie poster, --ar 16:9 --v 5.1

## 皮克斯　Pixar style

prompt: Pixar style , a young girl in a forest, 3D, surrounded by vibrant colors, with an airy atmosphere, warm lighting, and sharp details --ar 16:9 --v 5.1

 ## 賽博龐克　Cyberpunk

prompt: Cyberpunk , cityscape, neon lights, cyberpunk characters, vivid colors, futuristic feel --ar 16:9 --v 5.1

## 黑色低俗　Pulp noir

prompt: Pulp noir, femme fatale, red lipstick, fur coat, long gloves, diamond necklace, smoky jazz club, sultry atmosphere, mysterious man watching from afar, vintage microphone on stage, film noir style with low key lighting and shadows --ar 16:9 --v 5.1

## 好萊塢風格　Hollywood style

prompt: Hollywood style, glamorous actress, flowing red carpet gown, sparkling diamonds, classic updo hairstyle, dramatic makeup,  vintage film camera aesthetic, black and white color grading, film grain texture,l --ar 16:9 --v 5.1

蒙太奇　Montage

prompt: Montage, vintage photos, faded edges, overlapping images, nostalgic atmosphere, sepia tones, handwritten notes and scribbles, old film camera aesthetic, inspired by Robert Frank --ar 16:9 --v 5.1

 底片攝影風格　Film photography

prompt: Film photography, mysterious woman, soft focus, grainy texture, black and white film, dramatic lighting, moody atmosphere, abandoned building, tilted composition, by Alfred Hitchcock --ar 16:9 --v 5.1

## 微縮模型電影風格　Miniature movie style

prompt: Miniature movie style, tiny figurines, intricate details, dramatic lighting, moody atmosphere, vintage color grading, cinematic composition, sweeping camera movements, intense close-ups --ar 16:9 --v 5.1

## 巴洛克藝術　Baroque

prompt: Baroque, flamboyant and theatrical, masked ballroom scene with masquerade costumes and masks in rich fabrics and feathers, candlelit ambiance with shadows, elaborate hairstyles and jewelry for both men and women --ar 16:9 --v 5.1

## 法國藝術　French art

prompt: French art, soft pastel colors, impressionist brushstrokes, flowing lines, delicate details, outdoor garden scene, natural light, loose and free composition, dreamy atmosphere, by Claude Monet --ar 16:9 --v 5.1

## 包浩斯　Bauhaus

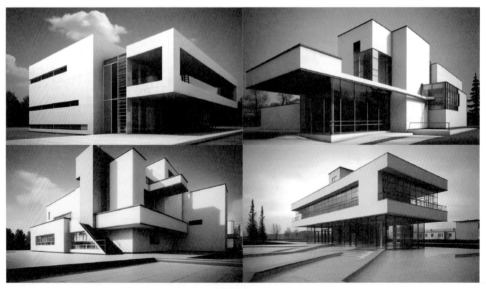

prompt: Bauhaus, clean lines and geometric shapes, monochromatic color scheme, large windows and natural light, asymmetrical composition, industrial materials like concrete and steel, futuristic atmosphere, by Walter Gropius --ar 16:9 --v 5.1

## 達文西　Leonardo Da Vinci

prompt: Leonardo Da Vinci, flowing beard, furrowed brow, quill pen in hand, surrounded by sketches and blueprints, candlelight casting dramatic shadows on his face, oil painting style with warm tones and rich textures --ar 16:9 --v 5.1

## 梵谷　Vincent Van Gogh

prompt: Vincent Van Gogh, starry night sky, swirling clouds, crescent moon, cypress trees, small village in the distance, bold brushstrokes, rich blues and yellows, dreamlike atmosphere, in expressionist style --ar 16:9 --v 5.1

## 克洛德・莫內　Claude Monet

prompt: Claude Monet, impressionist painting, vibrant colors, loose brushstrokes, water lilies in the background, natural light, outdoor setting, capturing the essence of nature and light, in the style of Monet himself --ar 16:9 --v 5.1

## 新藝術風格　Art nouveau

prompt: Art nouveau, intricate and ornate details, gilded surfaces, jewel tones, natural motifs, harmonious composition, romantic atmosphere, Alphonse Mucha-inspired style --ar 16:9 --v 5.1

## 洛可可　Rococo

prompt: Rococo, ornate and intricate gown with layers of ruffles and lace, powdered wig with jeweled hairpins, standing in a grand ballroom with crystal chandeliers and marble floors, surrounded by courtiers in elegant attire --ar 16:9 --v 5.1

## 文藝復興　Renaissance

prompt: Renaissance, majestic king, ornate golden crown, flowing robes, surrounded by courtiers and attendants, intricate details in the background, oil on canvas, chiaroscuro lighting, dramatic shadows and highlights --ar 16:9 --v 5.1

## 立體主義　Cubism

prompt: Cubism, distorted features and proportions, vibrant colors and bold brushstrokes, multiple perspectives and viewpoints, influenced by African masks and Picasso's style, on display in a retro art gallery with dim lighting and vintage furniture --ar 16:9 --v 5.1

## 印象派　Impressionistic

prompt: Impressionistic, portrait, vibrant colors, blurred features, dynamic brushstrokes, lively expression, energetic background, Gustav Klimt style --ar 16:9 --v 5.1

## 點描畫派　Pointillism

prompt: Pointillism, colorful dots, impressionist style, nature scene, trees and flowers, soft and dreamy atmosphere, Monet-inspired, oil painting texture, pastel colors, slightly blurred edges, asymmetrical composition --ar 16:9 --v 5.1

## 野獸派　Fauvism

prompt: Fauvism, vibrant colors, bold brushstrokes, expressive and exaggerated forms, landscape scene, sun-drenched countryside, lush greenery, warm hues of orange and yellow, contrasting with cool blues and purples --ar 16:9 --v 5.1

## 彩色玻璃窗　Stained glass window

prompt: Stained glass window, intricate geometric patterns, sunlight streaming through, casting vibrant hues on the surrounding walls and floor, Gothic architecture, cathedral setting, mystical and ethereal atmosphere --ar 16:9 --v 5.1

## 抽象風　Abstract

prompt: Abstract, fluid and organic shapes, vibrant colors, swirling and interlocking patterns, dynamic composition, ethereal atmosphere, reminiscent of a dream world, in digital art style, using Adobe Photoshop and Illustrator --ar 16:9 --v 5.1

孔版印刷風　Risograph

prompt: Risograph, vibrant and contrasting colors, organic shapes and lines, overlapping layers, textured paper, playful and whimsical atmosphere, in a square format, by Yayoi Kusama --ar 16:9 --v 5.1

 設計風　Graphic

prompt: Graphic, colorful abstract shapes, overlapping and interweaving, creating a sense of movement and energy, against a dark background, with a neon glow, in a digital art style, reminiscent of 80s arcade games contrast --ar 16:9 --v 5.1

**歐普藝術　OP art**

prompt: OP art, vibrant colors, swirling shapes, hypnotic patterns, kaleidoscopic effect, contrasting hues, dynamic composition, psychedelic atmosphere, neon lights, futuristic style --ar 16:9 --v 5.1

 **光效應藝術　Optical art**

prompt: Optical art, vibrant colors, overlapping circles and squares, op art illusion, dynamic movement, contrasted background, asymmetrical composition, modern and sleek style, in digital painting --ar 16:9 --v 5.1

## 次表面散射　Subsurface scattering

prompt: Subsurface scattering, ethereal and otherworldly, glowing from within, translucent skin, iridescent colors, floating in a misty atmosphere, delicate and graceful movements, soft and dreamy composition --ar 16:9 --v 5.1

 ## 蒸汽龐克　Steampunk

prompt: Steampunk, rusty gears and cogs, brass and copper, leather straps and buckles, Victorian era clothing, top hat and goggles, industrial machinery in the background, smoky atmosphere, gritty texture, by Tim Burton --ar 16:9 --v 5.1

## 普普藝術　Pop art

prompt: Pop art, bold and vibrant colors, comic book style, exaggerated features and expressions, halftone dots, speech bubbles, onomatopoeia, retro 60s vibe, Andy Warhol inspired --ar 16:9 --v 5.1

## 寫實主義　Realism

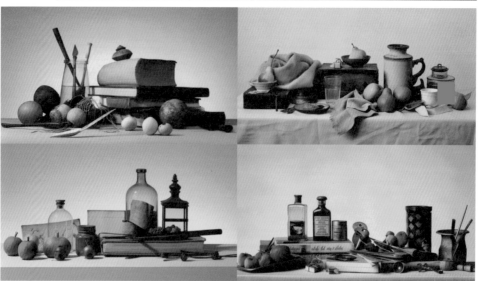

prompt: Realism, everyday objects arranged on a table, soft shadows, warm color tones, shallow depth of field, natural composition, vintage vibe --ar 16:9 --v 5.1

## 抽象表現主義　Abstract art

prompt: Abstract art, vibrant colors, bold brushstrokes, chaotic composition, energetic and dynamic atmosphere, mixed media materials, splatters and drips, with a sense of movement and flow --ar 16:9 --v 5.1

 ## 超寫實主義　Hyperrealistic

prompt: Hyperrealistic, intense gaze, sweat beads on forehead, wrinkles around the eyes and mouth, pores visible on skin, high contrast lighting, black and white, charcoal drawing style, by Chuck Close --ar 16:9 --v 5.1

超現實主義　Surrealism

prompt: Surrealism, morphing shapes and objects, bizarre creatures and landscapes, dream logic narrative, psychedelic color palette, stop-motion technique mixed with digital animation tools and software --ar 16:9 --v 5.1

 極簡主義　Minimalist

prompt: Minimalist, monochromatic color scheme, clean lines, geometric shapes, natural lighting, simple composition, negative space, serene atmosphere, digital art style --ar 16:9 --v 5.1

## 歌德式　Gothic style

prompt: Gothic style, black lace and velvet dress, intricate details, flowing cape, dark and moody lighting, candlelit room, ornate chandelier, dramatic makeup, pale skin, red lips, standing in front of a stained glass window, in a painterly style --ar 16:9 --v 5.1

## 歌德式黑暗　Gothic gloomy

prompt: Gothic gloomy, abandoned castle, broken windows, moss-covered walls, full moon in the sky, silhouette of a lone figure standing on the balcony, wearing a hooded cloak, ominous and foreboding atmosphere, in a dark fantasy style --ar 16:9 --v 5.1

粗獷主義　Brutalist

prompt: Brutalist, concrete monoliths, sharp edges and angles, minimalistic design, shadows and light play, stark and imposing, industrial feel, black and white photography, high contrast, symmetrical composition, by Julius Shulman --ar 16:9 --v 5.1

 建構主義　Constructivism

prompt: Constructivism, bold geometric shapes, primary colors, stark contrast, industrial setting, metal beams and concrete walls, sharp angles and lines, futuristic vibe, high-tech equipment in the background, in a digital art style with glitch effects --ar 16:9 --v 5.1

## 未來主義　Futuristic

prompt: Futuristic, towering skyscrapers with holographic projections, neon lights, flying cars and drones, hovering billboards, bustling crowds, chrome and glass architecture, sleek and minimalist design, high-tech gadgets and devices --ar 16:9 --v 5.1

## 工業風格　Industrial

prompt: Industrial, rusted metal pipes, steam rising, heavy machinery, sparks flying, workers in protective gear, gritty and dirty atmosphere, monochromatic color scheme with pops of bright yellow caution signs --ar 16:9 --v 5.1

書頁　Book page

prompt: Book page, yellowed and worn with age, intricate handwritten script in black ink, faded illustrations of flowers and leaves, curled edges and creases, dimly lit by a candle in an old-fashioned holder, in a rustic and cozy atmosphere --ar 16:9 --v 5.1

 雕刻藝術風格　Carving art

prompt: Carving art, intricate wood carving, delicate and ornate details, nature-inspired motifs, warm and earthy tones, soft lighting, close-up shot, shallow depth of field, emphasizing texture and depth --ar 16:9 --v 5.1

美式鄉村風格　American country

prompt: American country, wooden barn with red paint peeling off, tall grasses swaying in the wind, a vintage tractor parked nearby, blue skies with clouds, warm sunlight filtering through the trees, painted in oil on canvas with impressionist style --ar 16:9 --v 5.1

 真實的　Realistic

prompt: Realistic, fresh fruits and vegetables, wooden table as the background, natural light coming from the window on the right side, shallow depth of field to highlight the details of each piece of produce --ar 16:9 --v 5.1

## 3D 風格　3D

prompt: 3D, fantasy world, enchanted forest with towering trees and magical creatures, majestic castles and sprawling kingdoms, intricate details and ornate designs, vibrant colors and whimsical elements, in a high fantasy style --ar 16:9 --v 5.1

## 複雜的　Sophisticated

prompt: Sophisticated, intricate painting, with fine brushstrokes and intricate details, featuring a beautiful woman in a flowing dress, surrounded by flowers and butterflies, painted in oil on canvas --ar 16:9 --v 5.1

## 真實感　Photoreal

prompt: Photoreal, beautiful sunset, golden hour, palm trees silhouetted against the sky, warm tones, serene atmosphere, gentle waves lapping at the shore, seagulls flying in the distance --ar 16:9 --v 5.1

 ## 國家地理　National geographic

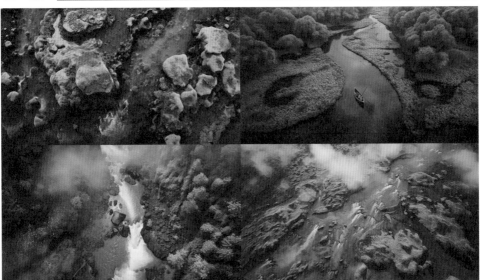

prompt: National geographic, breathtaking natural landscapes, vibrant colors, dynamic composition, wildlife in action, aerial shots, close-up details, adventurous spirit, captured by a drone, in high definition style --ar 16:9 --v 5.1

## 電影般的　Cinematic

prompt: Cinematic, post-apocalyptic world, deserted cityscapes, overgrown with vegetation, abandoned buildings, eerie silence, hauntingly beautiful music by Max Richter, desaturated and muted colors, gritty and dystopian style --ar 16:9 --v 5.1

## 建築素描　Architectural sketching

prompt: Architectural sketching, ornate details, intricate carvings and moldings, warm color palette, symmetrical composition, grandeur atmosphere, natural elements such as plants and water features incorporated into the design --ar 16:9 --v 5.1

**對稱肖像　Symmetrical portrait**

prompt: Symmetrical portrait, monochromatic color scheme, minimalist composition, high contrast lighting, sharp and defined features, geometric shapes in the background, in a surrealistic style --ar 16:9 --v 5.1

 **清晰的面部特徵　Clear facial features**

prompt: Clear facial features, soft and natural makeup, flowing hair, delicate jewelry, pastel colors, floral background, dreamy atmosphere, romantic style, shot with a Nikon D850 camera, 85mm lens, shallow depth of field --ar 16:9 --v 5.1

## 局部解剖　Partial anatomy

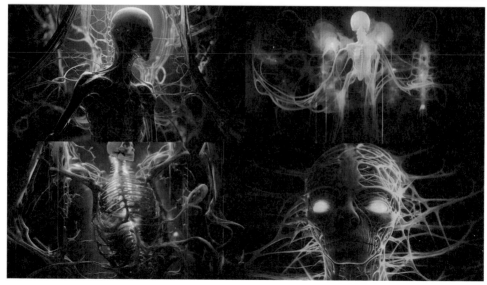

prompt: Partial anatomy, delicate veins and arteries, glowing neon green, floating in a dark void, surreal and dreamlike atmosphere, digital painting style with brush strokes visible --ar 16:9 --v 5.1

## 紡縫藝術　Quilted art

prompt: Quilted art, vibrant patchwork, geometric shapes and patterns, bold colors, layered textures, abstract composition, playful and whimsical atmosphere, mixed media collage style --ar 16:9 --v 5.1

## 復古黑暗　Retro dark vintage

prompt: Retro dark vintage, antique clock tower, towering over a small town square, gas lamps flickering in the night, misty and mysterious atmosphere, intricate details on the clock, composition --ar 16:9 --v 5.1

 ## 90 年代電玩遊戲　90s video game

prompt: 90s video game, pixelated graphics, neon colors, side-scrolling platformer, protagonist with spiky hair and oversized sword, enemies include robotic creatures and aliens, reminiscent of classic games like Super Mario --ar 16:9 --v 5.1

## 古典風格　Classical style

prompt: Classical style, vintage, antique furniture, ornate details, muted color palette, soft lighting, old books and paintings, elegant and sophisticated atmosphere --ar 16:9 --v 5.1

 ## 維多利亞時代　Victorian

prompt: Victorian, elegant lady, lace parasol, flowing dress, intricate embroidery, high collar, delicate gloves, walking in a garden, surrounded by blooming flowers and greenery, soft sunlight filtering through the trees --ar 16:9 --v 5.1

## 遊戲風格　Game style

prompt: Game style, retro pin-up girl, polka dot dress and headscarf, red lipstick and nails, vintage car with chrome accents, sitting on the hood with a coy expression, palm trees and sunny beach in the background --ar 16:9 --v 5.1

 ## 神秘幻想　Mystic fantasy

prompt: Mystic fantasy, ethereal and otherworldly, floating islands in the sky, cascading waterfalls, iridescent flora and fauna, a magical unicorn with a rainbow mane and tail, soaring through the clouds, in a dreamlike watercolor style --ar 16:9 --v 5.1

魔幻現實主義風格　Magic realism style

prompt: Magic realism style, floating islands in the sky, ancient ruins and temples, glowing runes and symbols, mysterious figures in hooded cloaks, otherworldly atmosphere, digital art style with neon accents --ar 16:9 --v 5.1

 角色概念藝術　Character concept art

prompt: Character concept art, fierce warrior princess, adorned with intricate gold armor, wielding a double-edged sword, standing atop a mountain peak overlooking a vast kingdom, surrounded by a stormy sky and bolts of lightning --ar 16:9 --v 5.1

遊戲場景圖　Game scene graph

prompt: Game scene graph, futuristic, neon lights, holographic displays, sleek metallic surfaces, floating platforms, advanced technology, high-tech weapons, intense action, cyberpunk atmosphere, in a stylized 3D render style --ar 16:9 --v 5.1

 武器設計　Weapon design

prompt: Weapon design, sleek and streamlined design, metallic silver finish, glowing blue accents, modular attachments for customization, energy-based ammunition, ergonomic grip and trigger, holographic sight --ar 16:9 --v 5.1

曠野之息　BOTW

prompt: BOTW, fantasy adventure, brave hero with a sword and shield, traveling through lush forests and treacherous mountains, epic battles against evil sorcerers and dark armies --ar 16:9 --v 5.1

 戰甲神兵　Warframe

prompt: Warframe, sleek and metallic armor, glowing energy blades, hovering jetpack, intense battle stance, neon-lit sci-fi setting, towering skyscrapers in the background --ar 16:9 --v 5.1

## 寶可夢　Pokémon

prompt: Pokémon, cute and fluffy, with big round eyes and rosy cheeks, surrounded by flowers and butterflies, sitting on a grassy meadow, in a watercolor style with soft colors --ar 16:9 --v 5.1

## Apex 英雄　APEX

prompt: APEX, colorful characters, each with unique abilities and weapons, standing on a futuristic platform overlooking a cityscape, ready for battle, dynamic poses and expressions, neon lights illuminating the scene --ar 16:9 --v 5.1

上古卷軸　The Elder Scrolls

prompt: The Elder Scrolls, ancient and mystical artifacts, glowing with power, hidden in a secret chamber, set in a fantasy world filled with dragons, wizards, and mythical creatures --ar 16:9 --v 5.1

 魂系遊戲　From software

prompt: From software, mysterious character, dark robes, glowing eyes, floating in mid-air, surrounded by ancient ruins and symbols, ominous atmosphere, monochromatic color scheme of blues and purples --ar 16:9 --v 5.1

## 底特律：變人　Detroit: Become Human

prompt: Detroit: Become Human, abandoned factory, rusted metal, broken machinery, neon lights, oppressive atmosphere, torn and dirty clothing --ar 16:9 --v 5.1

 ## 劍與遠征　AFK Arena

prompt: AFK Arena, hero characters, vibrant colors, fantastical creatures, intricate armor and weapons, epic battles, mystical landscapes, inspired by medieval art style, dynamic composition --ar 16:9 --v 5.1

跑跑薑餅人 Running Gingerbread Man

prompt: Running Gingerbread Man, whimsical cookie kingdom, candy-colored landscapes, towering gingerbread castles, swirling candy clouds, lollipop trees, in a dreamy and surreal style --ar 16:9 --v 5.1

 英雄聯盟 League of Legends

prompt: League of Legends, fierce battle scene, champions fighting with magic and weapons, intense energy and power, in a dark and ominous atmosphere, with a touch of fantasy and mythology --ar 16:9 --v 5.1

# 色彩質感

第**4**章

色彩質感是通過色彩的表現和組合傳達的視覺上的觸感或質感。它涉及人們對色彩的感知和情感反應，以及色彩所呈現的質地、光澤和觸感，可以通過調節色彩的明暗、飽和度、對比度和色調等要素來傳達。色彩質感的感知是主觀的，與個體的審美和情感密切相關。

紅　Red

prompt: Red, a field of blooming roses, close-up shot of a single rose with dew drops on its petals, soft and romantic atmosphere --ar 16:9 --v 5.1

 橙　Orange

prompt: Orange, dripping with sweet nectar, vibrant and fresh, surrounded by green leaves and sunlight streaming in through a nearby window, giving off a cozy and inviting atmosphere --ar 16:9 --v 5.1

## 黃 Yellow

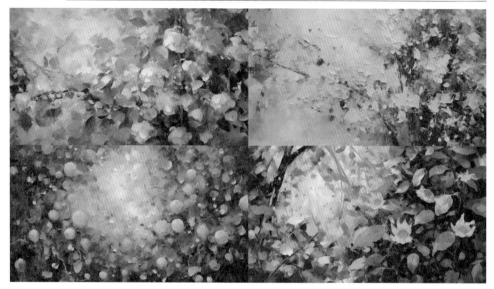

prompt: Yellow, round shape, citrus scent, warm and inviting atmosphere, impressionist style with visible brushstrokes and vibrant colors --ar 16:9 --v 5.1

## 綠 Green

prompt: Green, lush forest, towering trees, dappled sunlight, moss-covered rocks and fallen logs, chirping birds and rustling leaves, misty atmosphere, peaceful and serene, watercolor painting style --ar 16:9 --v 5.1

## 青 Cyan

prompt: Cyan, mystical forest, ancient trees with gnarled roots and twisted branches, misty atmosphere, glowing mushrooms, magical flowers with healing properties, in a dreamlike watercolor style --ar 16:9 --v 5.1

## 藍 Blue

prompt: Blue, vast ocean view, white sand beach, palm trees swaying in the breeze, clear blue sky, distant sailboats, seagulls soaring overhead, peaceful and serene atmosphere, impressionist painting style --ar 16:9 --v 5.1

## 紫 Purple

prompt: Purple, vibrant and energetic, neon lights illuminating a bustling city street, people dancing and laughing in a club,  bold and modern style --ar 16:9 --v 5.1

## 灰 Gray

prompt: Gray, living room, rustic decor, plush sofa with throw pillows, warm blankets and soft rugs, flickering fireplace, shelves lined with books and family photos --ar 16:9 --v 5.1

## 棕 Brown

prompt: Brown, leather suitcase, adorned with travel stickers from all over the world, sitting on a wooden floor next to an old globe and a stack of books --ar 16:9 --v 5.1

## 白 White

prompt: White, winter wonderland, snow-covered trees and mountains, icy blue streams, cozy cabin with smoke coming out of the chimney --ar 16:9 --v 5.1

## 黑 Black

prompt: Black, sports car, glossy finish, sharp edges, aerodynamic design, roaring engine sound, speeding down an empty highway at night, --ar 16:9 --v 5.1

## 薄荷綠色系 Mint green

prompt: Mint green, soft and delicate, pastel color palette, floral patterns, vintage aesthetic, flowing fabric, natural lighting, romantic atmosphere, watercolor painting style --ar 16:9 --v 5.1

## 日暮色系　Sunset gradient

prompt: Sunset gradient, bold and vibrant colors, dynamic and energetic atmosphere, city skyline silhouette in the foreground, contrasting against the warm hues of the sky --ar 16:9 --v 5.1

## 楓葉紅色系　Maple red

prompt: Maple red, floating on a tranquil lake, golden sunset reflecting on the water, misty mountains in the background, warm colors, natural composition --ar 16:9 --v 5.1

## 雪山藍色系　Mountain blue

prompt: Mountain blue, jagged peaks piercing the sky, snow-capped and glistening in the sunlight, icy blue glaciers carving through the valleys, a lone eagle soaring overhead --ar 16:9 --v 5.1

## 雷射糖果紙色　Laser candy paper color

prompt: Laser candy paper color, pastel tones, soft and fluffy texture, arranged in a heart shape, surrounded by glittering stars and sparkles, in a dreamy and whimsical setting, watercolor technique --ar 16:9 --v 5.1

## 馬卡龍色　Macarons

prompt: Macarons, delicate and smooth texture, various flavors such as lavender, rose, and pistachio, arranged in a circular pattern on a white plate --ar 16:9 --v 5.1

## 莫蘭迪色系　Muted tones

prompt: Muted tones, serene landscape, misty mountains and forests, calm waters and reflections, solitary figure in the distance, natural color palette --ar 16:9 --v 5.1

## 鈦金屬色系　Titanium

prompt: Titanium, sleek and modern design, sharp edges and angles, minimalist style, high shine finish, contrasting black accents, glowing blue lights --ar 16:9 --v 5.1

## 鮮果色系　Fresh fruits

prompt: Fresh fruits, vibrant colors, scattered on a rustic wooden table, natural lighting, water droplets glistening on the surface, organic shapes and textures, mixed varieties of apples, pears, and oranges --ar 16:9 --v 5.1

黑白灰色系　Black and white

prompt: Black and white, intense emotions, raw and unfiltered expression, close-up shot,  natural lighting, candid moment captured in time, rough texture of skin and hair --ar 16:9 --v 5.1

 極簡黑白色系　Minimalist black and white

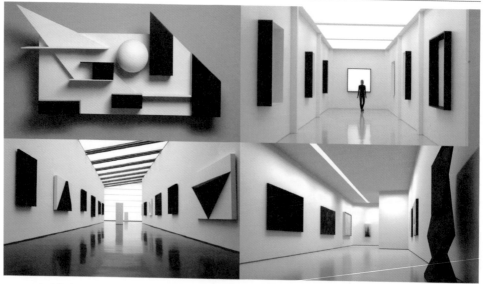

prompt: Minimalist black and white, geometric shapes, sharp lines, high contrast, abstract composition, monochromatic palette, strong sense of simplicity and elegance --ar 16:9 --v 5.1

## 溫暖棕色系　Warm brown

prompt: Warm brown, cozy knit sweater, cable knit pattern, surrounded by rustic wooden decor and soft blankets, natural light streaming in through the windows --ar 16:9 --v 5.1

 ## 柔和粉色系　Soft pink

prompt: Soft pink, fluffy clouds, cotton candy texture, pastel shades, dreamy atmosphere, ethereal beauty, watercolor painting style, floating in a peaceful sky --ar 16:9 --v 5.1

## 水晶藍色系　Crystal blue

prompt: Crystal blue, shimmering like diamonds, reflecting the sky, surrounded by lush greenery, gentle breeze blowing, birds chirping --ar 16:9 --v 5.1

## 發光　Shine

prompt: Shine, gleaming diamond, multi-faceted and reflective, against a black velvet background, highlighting the intricate details of the diamond and the craftsmanship of the jewelry piece --ar 16:9 --v 5.1

## 星閃　Star flash

prompt: Star flash, cosmic explosion, glittering stars, swirling galaxies, vibrant energy, dynamic movement, kaleidoscopic patterns, futuristic style, in 3D animation --ar 16:9 --v 5.1

## 螢光　Fluorescence

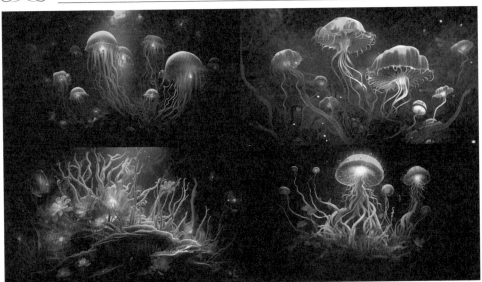

prompt: Fluorescence, bioluminescent creatures and plants, shimmering light rays and bubbles, deep blue hues and gradients, in a surrealistic and painterly style --ar 16:9 --v 5.1

## 聖光　Holy light

prompt: Holy light, intense and powerful, shining down on a majestic cathedral, illuminating intricate stained glass windows, casting vibrant colors across the interior, --ar 16:9 --v 5.1

## 反射透明彩虹色　Reflections transparent iridescent colors

prompt: Reflections transparent iridescent colors, iridescent colors, reflections on a car's surface, metallic shine, sleek and modern design, futuristic atmosphere, sharp lines and angles --ar 16:9 --v 5.1

**糖果色系　Candy**

prompt: Candy, rainbow colors, sugar crystals sparkling on the surface, round and smooth texture, surrounded by colorful confetti --ar 16:9 --v 5.1

 **珊瑚色系　Coral**

prompt: Coral, vibrant colors, diverse marine life, schools of fish, sea turtles, octopus, intricate textures and patterns, crystal clear water --ar 16:9 --v 5.1

薰衣草色系　Lavender

prompt: Lavender, macarons, delicate and airy texture, light purple color, next to a cup of tea and a bouquet of fresh lavender, in a cozy French patisserie --ar 16:9 --v 5.1

 綠寶石色系　Emerald

prompt: Emerald, green velvet couch, plush and luxurious, adorned with golden tassels and fringes, surrounded by antique furniture and decor, captured in a vintage film noir style --ar 16:9 --v 5.1

## 玫瑰金色系　Rose gold

prompt: Rose gold, luxurious and ornate wedding cake, intricate floral designs with edible gold leaf accents, multiple tiers with cascading sugar flowers --ar 16:9 --v 5.1

 ## 淺藍色系　Sky blue

prompt: Sky blue, fluffy clouds, serene atmosphere, birds flying freely, gentle breeze, watercolor painting style, impressionist brushstrokes, soft edges --ar 16:9 --v 5.1

## 酒紅色系　Burgundy

prompt: Burgundy, elegant and sophisticated, flowing fabric, long sleeves, fitted waist, sweeping train, luxurious ballroom setting, vintage glamour style --ar 16:9 --v 5.1

## 藍綠色系　Turquoise

prompt: Turquoise, peacock feathers, iridescent and shimmering in the light, arranged in a dramatic fan shape, surrounded by lush green foliage, exotic and vibrant atmosphere --ar 16:9 --v 5.1

## 黑色背景居中　Black background centered

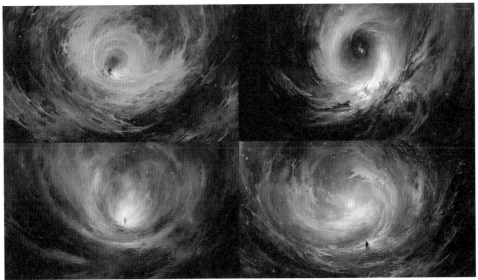

prompt: Black background centered, black hole, swirling cosmic dust and gas, glowing bright blue and white, massive and powerful gravitational pull, surrounded by distant stars and galaxies --ar 16:9 --v 5.1

## 白色和綠色調　White and green tones

prompt: White and green tones, ethereal forest, towering trees with moss-covered trunks, delicate white flowers blooming on the ground --ar 16:9 --v 5.1

## 紅色和黑色調　Red and black tones

prompt: Red and black tones, fierce warrior, adorned in red and black armor, wielding a double-bladed sword, standing on a rocky terrain with a stormy sky in the background --ar 16:9 --v 5.1

## 黃色和黑色調　Yellow and black tones

prompt: Yellow and black tones, fierce tiger, bold stripes, piercing eyes, powerful stance, jungle scene, misty atmosphere, water droplets on fur --ar 16:9 --v 5.1

## 金色和銀色調　Gold and silver tones

prompt: Gold and silver tones, jewelry, intricate filigree designs, sparkling diamonds and gemstones, elegant and refined, displayed on a velvet cushion in a glass case --ar 16:9 --v 5.1

 ## 霓虹色調　Neon shades

prompt: Neon shades, futuristic cityscape, towering skyscrapers and high-tech buildings, holographic advertisements and signs --ar 16:9 --v 5.1

## 亮麗橙色系　Bright orange

prompt: Bright orange, juicy and plump, citrusy scent, tropical paradise, wearing a straw hat and oversized sunglasses, relaxing on a pink flamingo pool float --ar 16:9 --v 5.1

## 象牙白色系　Ivory white

prompt: Ivory white, delicate lace, flowing fabric, ethereal atmosphere, soft lighting, vintage setting, antique furniture, dreamy and romantic style --ar 16:9 --v 5.1

## 自然綠色系　Natural green

prompt: Natural green, lush greenery, verdant leaves, sunlight filtering through the canopy, a babbling brook running through the scene, moss-covered rocks lining the stream --ar 16:9 --v 5.1

## 奢華金色系　Luxurious gold

prompt: Luxurious gold, intricate filigree patterns, shimmering and sparkling, lavish and extravagant, regal crown, ornate details, glowing with radiance, --ar 16:9 --v 5.1

## 穩重藍色系　Steady blue

prompt: Steady blue, ocean, calm and peaceful, gentle waves, clear sky, white sand beach, palm trees swaying in the breeze, a lone sailboat on the horizon --ar 16:9 --v 5.1

 ## 經典紅黑白色系　Classic red black and white

prompt: Classic red black and white, bold geometric shapes, minimalist style, high contrast, sharp edges, clean lines, abstract composition, --ar 16:9 --v 5.1

## 珊瑚橙色系　Coral orange

prompt: Coral orange, soft and fluffy texture, underwater scene, schools of vibrant fish, graceful sea turtles, warm and inviting sunlight filtering through the water --ar 16:9 --v 5.1

 ## 巨無霸色系　Whopper

prompt: Whopper, perfectly grilled beef patty, melted cheddar cheese, crispy lettuce, ripe tomatoes, sweet and savory special sauce, sesame seed bun, served on a red and white checkered paper in a classic American diner setting --ar 16:9 --v 5.1

## 秋日棕色系　Autumn brown

prompt: Autumn brown, cozy sweater, woolen texture, sitting by the fireplace, reading a book, surrounded by autumn leaves, warm and inviting atmosphere --ar 16:9 --v 5.1

## 丹寧藍色系　Denim blue

prompt: Denim blue, faded and distressed, patchwork design, high-waisted jeans, cropped length, frayed hem, oversized white shirt tucked in, vintage sunglasses --ar 16:9 --v 5.1

## 時尚灰色系　Fashionable gray

prompt: Fashionable gray, sleek and modern suit, tailored fit, minimalistic design, silver accessories, cool blue tones, dramatic lighting, cinematic composition --ar 16:9 --v 5.1

## 芭比粉色系　Barbie pink

prompt: Barbie pink, vintage and retro, polka dots and stripes, flared skirt and cat-eye sunglasses, classic car as background, palm trees and blue sky, 1950s Miami Beach vibe --ar 16:9 --v 5.1

## 紫羅蘭紫色系　Violet purple

prompt: Violet purple, delicate flowers, soft petals, dew drops, green leaves, natural lighting --ar 16:9 --v 5.1

## 彩虹色系　Iridescence

prompt: Iridescence, vibrant rainbow, arcing across a clear blue sky, radiating colors of red, orange, yellow, green, blue, indigo and violet --ar 16:9 --v 5.1

## 啞光質感　Matte texture

prompt: Matte texture, monochromatic tones, smooth and velvety surface, subtle variations in shade, simple geometric shapes --ar 16:9 --v 5.1

## 珍珠質感　Pearl texture

prompt: Pearl texture, nestled in an oyster shell, iridescent shades of pink and blue, delicate ridges and curves, surrounded by shimmering water droplets --ar 16:9 --v 5.1

## 綢緞質感　Silk texture

prompt: Silk texture, flowing and delicate, iridescent colors shimmering in the light, billowing like waves, soft to the touch --ar 16:9 --v 5.1

## 毛絨質感　Fluffy texture

prompt: Fluffy texture, soft and cloud-like, pastel colors, dreamy and ethereal atmosphere, close-up shot, macro lens, shallow depth of field, bokeh effect --ar 16:9 --v 5.1

## 水波紋質感　Water wave texture

prompt: Water wave texture, gentle and flowing, with white foam and bubbles, sun-kissed sparkles on the surface, in a minimalist style, using watercolor brush strokes --ar 16:9 --v 5.1

 ## 珠光質感　Pearl luster texture

prompt: Pearl luster texture, smooth and shiny surface, iridescent colors, floating in a dark abyss, surrounded by glowing jellyfish, ethereal and dreamlike atmosphere --ar 16:9 --v 5.1

## 竹子質感　Bamboo texture

prompt: Bamboo texture, bamboo forest, misty and serene, tall and slender bamboo stalks, sunlight filtering through the leaves, gentle breeze rustling the leaves, --ar 16:9 --v 5.1

## 金屬質感　Metallic texture

prompt: Metallic texture, sleek and shiny, reflecting rainbow hues, industrial and futuristic, sharp edges and smooth curves, abstract and geometric shapes, in monochromatic color scheme --ar 16:9 --v 5.1

## 石頭質感　Stone texture

prompt: Stone texture, jagged edges, irregular shapes, muted earthy tones, weathered appearance, mossy growth, natural lighting, in black and white style --ar 16:9 --v 5.1

 ## 石墨質感　Graphite texture

prompt: Graphite texture, rough and gritty, dark and moody, metallic accents, jagged edges, abstract shapes, monochromatic color scheme, in a minimalist and modern style --ar 16:9 --v 5.1

## 玻璃質感　Glass texture

prompt: Glass texture, crystal clear glass, smooth surface, refracting light, minimalist design, geometric shapes, clean lines --ar 16:9 --v 5.1

## 皮革質感　Leather texture

prompt: Leather texture, rich and supple, deep brown color, intricate stitching details, glossy finish, under warm and soft lighting --ar 16:9 --v 5.1

## 塑膠質感　Plastic texture

prompt: Plastic texture, glossy and reflective, smooth and seamless, bright neon colors, geometric shapes, abstract patterns --ar 16:9 --v 5.1

## 水晶質感　Crystal texture

prompt: Crystal texture, iridescent and shimmering, sharp edges and facets, refracting light in all directions, set against a dark background --ar 16:9 --v 5.1

## 棉質　Cotton texture

prompt: Cotton texture, fluffy and cozy, pastel colors, delicate embroidery, floral patterns, vintage style, in a quaint countryside cottage --ar 16:9 --v 5.1

## 沙質　Sandy texture

prompt: Sandy texture, rough and grainy, beige and tan tones, scattered seashells and driftwood, natural and organic feel, watercolor painting style --ar 16:9 --v 5.1

## 陶瓷質感　Ceramic texture

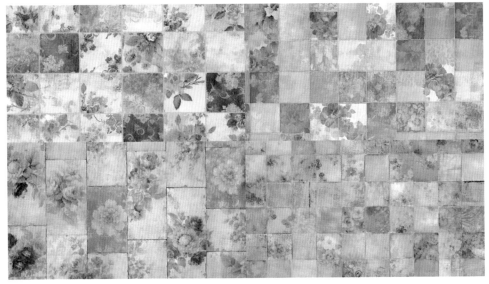

prompt: Ceramic texture, crackled and distressed finish, vintage floral patterns with faded colors of pink and blue, irregularly shaped tiles arranged in a patchwork pattern --ar 16:9 --v 5.1

## 紗網質感　Gauze texture

prompt: Gauze texture, ethereal beauty, flowing gauze dress, delicate lace details, soft pastel colors, dreamy atmosphere --ar 16:9 --v 5.1

## 磚石質感　Brick texture

prompt: Brick texture, rough and rustic, earthy tones, weathered appearance, irregular shapes and sizes, interlocking pattern, moss and vines growing in crevices --ar 16:9 --v 5.1

## 古銅質感　Antique bronze texture

prompt: Antique bronze texture, intricate carvings and details, ornate and elegant, aged patina, warm and inviting, set against a dark background --ar 16:9 --v 5.1

## 油漆質感　Varnish texture

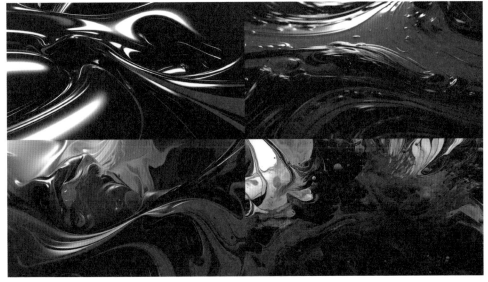

prompt: Varnish texture, glossy and reflective, deep red hue, smooth and polished surface, intricate swirls and patterns --ar 16:9 --v 5.1

## 金屬漆質感　Metallic paint texture

prompt: Metallic paint texture, abstract shapes, bold and vibrant colors, high contrast, glossy finish, dynamic composition,  3D modeling software, --ar 16:9 --v 5.1

## 菌絲　Mycelium

prompt: Mycelium, delicate and intricate branching patterns, neon colors, glowing in the dark, growing on decaying wood --ar 16:9 --v 5.1

## 木頭　Wood

prompt: Wood, rough and textured bark, twisted and gnarled branches, dappled sunlight filtering through the leaves, a bird perched on a branch --ar 16:9 --v 5.1

**腐朽衰敗的　Decayed**

prompt: Decayed, wilting flowers, peeling paint, rusted metal, chipped wood, faded colors, abandoned building, eerie atmosphere, vintage style --ar 16:9 --v 5.1

 **骨骼狀　Skeletal**

prompt: Skeletal, bone carving, intricate and delicate details, natural materials, organic shapes and textures, featuring animals and symbols with spiritual significance --ar 16:9 --v 5.1

## 玻璃　Glass

prompt: Glass, sharp edges, scattered pieces, reflecting light, abstract shapes, contrasting colors, dark background --ar 16:9 --v 5.1

## 棉花　Cotton

prompt: Cotton, endless rows of white fluff stretching towards the horizon, under a bright blue sky with fluffy clouds --ar 16:9 --v 5.1

亞麻布　Linen

prompt: Linen, soft and delicate texture, natural beige color, wrinkled and creased surface, draped over a wooden chair, captured in black and white photography style --ar 16:9 --v 5.1

 蕾絲　Lace

prompt: Lace, intricate floral patterns, soft and feminine, vintage feel, ivory color,　--ar 16:9 --v 5.1

## 瓷器 Porcelain

prompt: Porcelain, teapot, traditional Chinese design with intricate patterns and gold accents, steam rising from the spout --ar 16:9 --v 5.1

 青瓷 Celadon

prompt: Celadon, vase, intricate floral patterns, delicate curves, crackled glaze, sitting on a wooden pedestal, surrounded by lush greenery and natural ligh --ar 16:9 --v 5.1

## 琺瑯　Enamel

prompt: Enamel, glossy and smooth surface, vibrant colors, intricate designs, geometric shapes, floral patterns --ar 16:9 --v 5.1

## 黏土質感　Clay texture

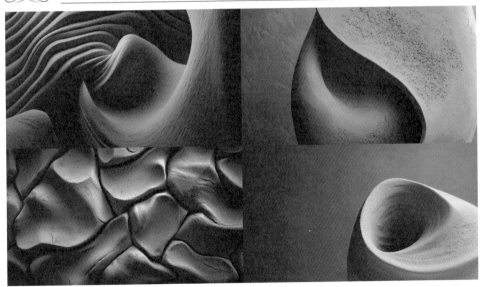

prompt: Clay texture, organic shapes, earthy tones, natural light source, subtle shadows, tactile feel, in monochromatic black and white style --ar 16:9 --v 5.1

## 紋理質感　Texture

prompt: Texture, rough and rugged surface, weathered and aged, with cracks and crevices, earthy tone --ar 16:9 --v 5.1

## 皮毛質感　Fur texture

prompt: Fur texture, glossy and reflective, shimmering with iridescent hues of green and blue, adorning the coat of a majestic panther, piercing yellow eyes glowing in the darkness --v 5.1

## 雕刻質感　Carved texture

prompt: Carved texture, rough and uneven surface, deep grooves and sharp edges, natural wood grain, warm and earthy tones, rustic and vintage feel --ar 16:9 --v 5.1

 ## 砂岩　Sandstone

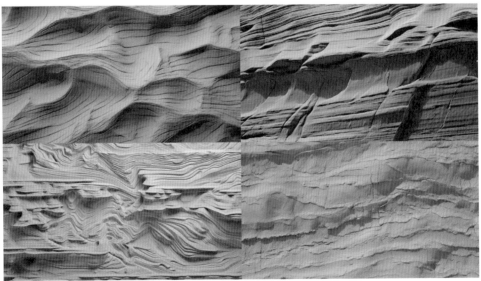

prompt: Sandstone, rough and textured surface, warm beige tone, eroded by wind and water, intricate patterns of sedimentary layers, contrast between smooth and rough textures --ar 16:9 --v 5.1

## 天鵝絨　Velvet

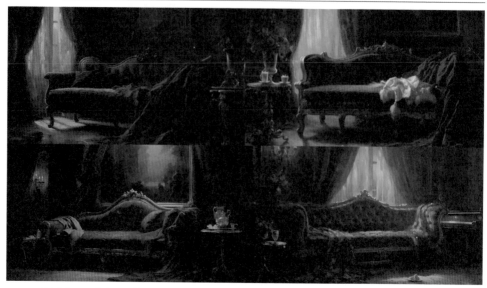

prompt: Velvet, rich and luxurious texture, deep red color, soft to the touch, draped over a chaise lounge, golden accents,vintage vibe --ar 16:9 --v 5.1

## 薄紙巾　Tissue paper

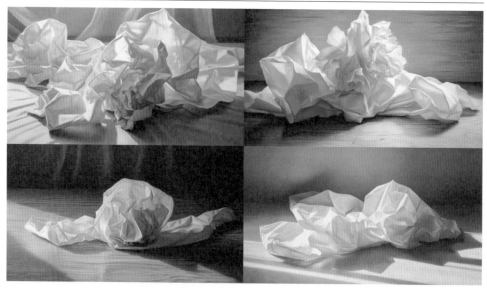

prompt: Tissue paper, soft and delicate texture, pastel colors, scattered on a wooden surface, natural light casting shadows --ar 16:9 --v 5.1

## 曲線細膩的　Delicately curved

prompt: Delicately curved, flowing lines, soft and smooth textures, pastel colors, dreamy atmosphere, natural setting, surrounded by flowers and foliage --ar 16:9 --v 5.1

## 有層次感的　Layered

prompt: Layered, intricate and delicate patterns, muted color palette, soft and flowing fabrics, multiple textures, ethereal atmosphereimpressionist style --ar 16:9 --v 5.1

## 線條優美且精細的　Exquisite

prompt: Exquisite, flower arrangement, delicate and colorful blooms, soft focus, vintage aesthetic, in a rustic pottery vase, captured in a dreamy watercolor style --ar 16:9 --v 5.1

 ## 線條簡潔且清晰的　Simple and clear lines

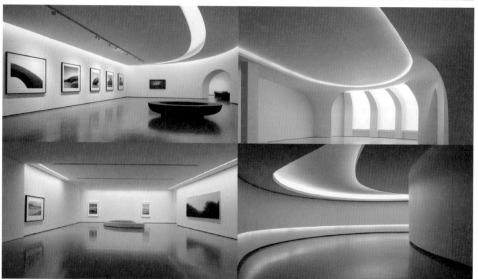

prompt: Simple and clear lines, elegant curves, flowing and graceful, smooth and polished surface, natural lighting, in a spacious and modern gallery setting --ar 16:9 --v 5.1

## 有紋理的　Textured

prompt: Textured, smooth and silky fabric, jewel tones, geometric shapes, dramatic lighting, asymmetrical composition, mysterious atmosphere, oil painting style --ar 16:9 --v 5.1

## 線條彎曲但流暢的　Sinuous, fluid curves

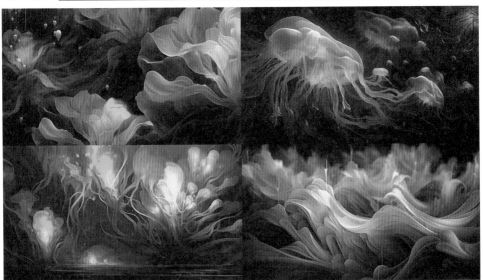

prompt: Sinuous, fluid curves, undulating waves, organic shapes, iridescent colors, underwater world, bioluminescent creatures, --ar 16:9 --v 5.1

## 紋路自然的　Organic pattern

prompt: Organic pattern, flowing and curving lines, earthy colors, reminiscent of nature, in a circular shape, surrounded by smaller patterns, in a hand-drawn style --ar 16:9 --v 5.1

## 大膽的顏色　Bold color

prompt: Bold color, abstract shapes, minimalistic composition, high contrast, sharp edges, geometric patterns, in digital art style --ar 16:9 --v 5.1

##  具有浮雕感的　Embossed

prompt: Embossed, bronze texture, floral motifs, romantic design, vintage feel, on a wooden jewelry box, with brass hardware and intricate carvings --ar 16:9 --v 5.1

##  具有雕刻感的　Carved

prompt: Carved, ice sculpture, delicate and intricate details, translucent texture, nature-inspired design with floral elements and animals, illuminated from within with colorful lights --ar 16:9 --v 5.1

## 超細節　Epic detail

prompt: Epic detail, mythical creature, iridescent scales, intricate patterns, glowing eyes, majestic wingspan,surrounded by mist and fog, oil painting style --ar 16:9 --v 5.1

## 光滑的　Smooth

prompt: Smooth, sleek and modern design, glossy surfaces, metallic accents, bold and vibrant colors, geometric shapes, high-tech materials --ar 16:9 --v 5.1

清晰的 Clear

prompt: Clear, water droplets, delicately hanging from the edge of a leaf, reflecting the vibrant green of the surrounding foliage, captured in macro photography style, --ar 16:9 --v 5.1

 細膩的 Delicate

prompt: Delicate, flower, soft petals, pastel colors, dew drops, morning light, close-up shot, shallow depth of field, bokeh background --ar 16:9 --v 5.1

## 精細的　Fine

prompt: Fine, dining, elegant table setting, intricate silverware and crystal glasses, colorful and fragrant dishes, artistic plating --ar 16:9 --v 5.1

## 平整的　Flat

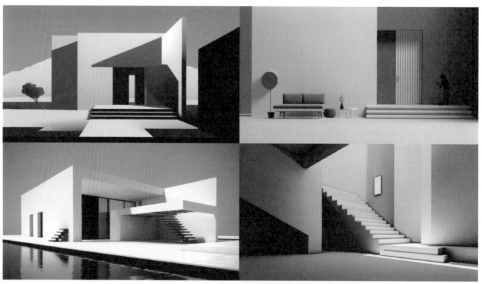

prompt: Flat, minimalist, monochromatic color scheme, clean lines, geometric shapes, natural light, uncluttered composition, in a modern and sleek style --ar 16:9 --v 5.1

## 精密的　Precise

prompt: Precise, measuring instruments, sleek metallic finish, digital readouts for accurate and reliable measurements, in a modern laboratory with advanced technology and equipment --ar 16:9 --v 5.1

## 線條流暢的　Sleek

prompt: Sleek, modern car, glossy black exterior, aerodynamic curves, tinted windows, chrome accents, LED headlights --ar 16:9 --v 5.1

**流線型的　Streamlined**

prompt: Streamlined, sleek and modern design, minimalist approach, metallic finish, sharp edges and curves, high-tech materials, futuristic atmosphere --ar 16:9 --v 5.1

 **線條優美的　Graceful**

prompt: Graceful, ballerina, black and white photography, strong contrast, dramatic lighting, pointe shoes, tutu skirt, elegant poses and movements --ar 16:9 --v 5.1

## 彎曲的　Curved

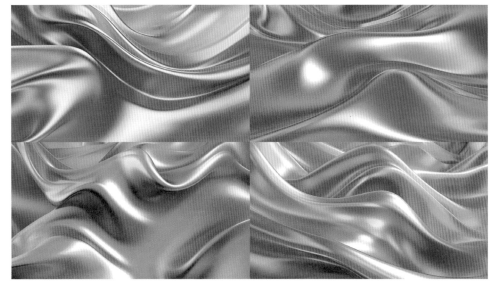

prompt: Curved, smooth surface, iridescent rainbow colors, reflective metallic texture, organic shape, undulating waves, soft and gentle lighting --ar 16:9 --v 5.1

## 柔和的　Soft

prompt: Soft, fluffy pillow, pastel pink and blue, overstuffed with down feathers, gentle curves and folds, placed on a cozy bed with a warm woolen blanket, surrounded by dimly lit candles --ar 16:9 --v 5.1

多樣化的　Varied

prompt: Varied, an eclectic mix of cultures and traditions, blending old and new, east and west, in a vibrant and dynamic scene, in a style that fuses different artistic influences --ar 16:9 --v 5.1

 有機的　Organic

prompt: Organic, lush farm, rows of leafy green vegetables, buzzing bees and fluttering butterflies, rustic wooden fence, sun-drenched fields --ar 16:9 --v 5.1

## 光潔的　Polished

prompt: Polished, glossy finish, light gray color, stacked in a modern and minimalistic style, in a sleek and contemporary office building, surrounded by glass walls and metal accent --ar 16:9 --v 5.1

## 微妙的　Subtle

prompt: Subtle, soft pastel colors, delicate flowers, dew drops, gentle breeze, natural light, bokeh effect, vintage film style, 35mm film camera, shallow depth of field --ar 16:9 --v 5.1

## 纖細的　Slender

prompt: Slender, dancer, wearing a shimmering silver leotard and pointe shoes, gracefully performing a ballet dance on stage under the spotlight --ar 16:9 --v 5.1

## 細長的　Thin

prompt: Thin, delicate and graceful, flowing dress, soft pastel colors, watercolor painting style --ar 16:9 --v 5.1

## 線條細致的　Intricate

prompt: Intricate, maze, twisting and turning, with hidden paths and dead ends, filled with mythical creatures and magical elements, rendered in a hand-drawn art style --ar 16:9 --v 5.1

## 線條流暢且柔和的　Gentle

prompt: Gentle, giant, towering over the landscape, kind eyes and a warm smile, covered in soft fur or feathers, surrounded by nature and wildlife --ar 16:9 --v 5.1

## 優雅的　Elegant

prompt: Elegant, a lady, flowing silk dress, pastel colors, delicate lace details, soft curls, holding a bouquet of white roses, standing in a blooming garden, natural light,  --ar 16:9 --v 5.1

## 曲線優美的　Curvaceous curves

prompt: Curvaceous curves, vintage dress with intricate lace details, sitting in a Victorian-style room, romantic and nostalgic atmosphere --ar 16:9 --v 5.1

## 療癒的　Healing

prompt: Healing, tranquil forest, misty morning, soft sunlight, gentle breeze, rustling leaves, towering trees, chirping birds, watercolor painting style --ar 16:9 --v 5.1

## 猛烈的　Rough

prompt: Rough, rugged terrain, steep cliffs, howling wind, stormy skies, lightning strikes, dramatic lighting, intense atmosphere, black and white photography style --ar 16:9 --v 5.1

## 不規則的　Irregular

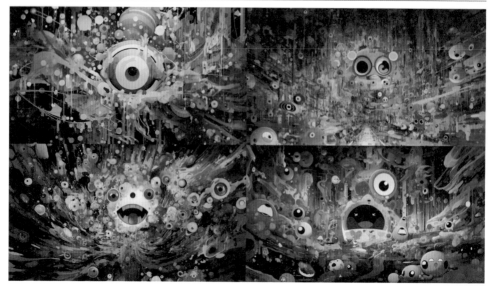

prompt: Irregular, abstract shapes, vibrant colors, chaotic composition, surreal atmosphere, digital art style,neon lights, kaleidoscopic patterns --ar 16:9 --v 5.1

## 龐大的　Bulky

prompt: Bulky, an elephant, long tusks, standing tall and proud, surrounded by lush greenery, peaceful and serene atmosphere, captured in watercolor painting style --v 5.1

## 鋭利的　Sharp

prompt: Sharp, crisp lines, minimalist design, monochromatic color scheme, geometric shapes, high contrast lighting, isolated subject on a white background --ar 16:9 --v 5.1

## 多稜角的　Angular

prompt: Angular, towering skyscrapers, glass facades, sharp angles and lines, reflecting the blue sky and white clouds --ar 16:9 --v 5.1

## 充滿動感的　Dynamic

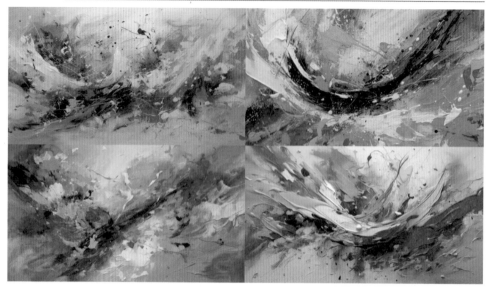

prompt: Dynamic, explosive energy, bold and vibrant colors, swirling shapes and lines, fast-paced movement, sense of motion blur, abstract expressionism, --ar 16:9 --v 5.1

## 統一的　Uniform

prompt: Uniform, crisp and clean, pressed collar, polished buttons, pleated skirt, knee-high socks, shiny black shoes --ar 16:9 --v 5.1

# 光照效果

第**5**章

營造光照效果對於創造逼真的場景和物體非常重要。
通過營造光照效果,我們能夠呈現出物體的明暗變化、
表面的紋理細節和陰影的深度與形狀。從柔和的環境
光到強烈的鏡面反射,每一種光照效果都對應著特殊
的畫面效果。

## 電影光　Cinematic light

prompt: Cinematic light, warm and soft glow, filtering through the trees, casting shadows on the ground, creating a peaceful and serene atmosphere, golden hour colors, lens flares and bokeh, nature-inspired composition, wide-angle shot, --ar 16:9 --v 5.1

## 戲劇燈光　Dramatic lighting

prompt: Dramatic lighting, moody and intense, deep shadows and bright highlights, contrasting colors, sharp and defined edges, mysterious atmosphere, film noir style, black and white, high contrast, low key lighting --ar 16:9 --v 5.1

## 強光逆光　Intense backlight

prompt: Intense backlight, silhouette of a lone figure, stark contrast, dramatic shadows, minimal details, monochromatic color scheme, high key lighting, mysterious atmosphere, minimalist composition, in black and white film noir style --ar 16:9 --v 5.1

## 立體光　Volumetric light

prompt: mystical forest, Volumetric light shining through the trees, majestic deer standing tall and proud, ancient ruins in the background, moss-covered stones and overgrown vines, a sense of magic and wonder permeating the scene, --ar 16:9 --v 5.1

## 閃光燈光　Shimmering light

prompt: Shimmering light, ethereal and delicate, dancing in the wind, reflecting off water droplets, soft pastel colors, dreamlike atmosphere, wispy clouds in the background, in a painterly style with brushstrokes visible, inspired by Monet --ar 16:9 --v 5.1

## 影棚光　Studio light

prompt: Studio light, soft and diffused, illuminating a vintage camera on a wooden table, casting shadows of the camera's intricate details onto the table, creating a warm and nostalgic atmosphere, in a classic black and white style --ar 16:9 --v 5.1

**雙性照明 Bisexual lighting**

prompt: Bisexual lighting, soft and warm hues of pink and blue, intermingling in a dreamy and ethereal way, casting a gentle glow on the subject, creating a romantic and intimate atmosphere, with a touch of mystery and intrigue --ar 16:9 --v 5.1

 **反射 Reflection effect**

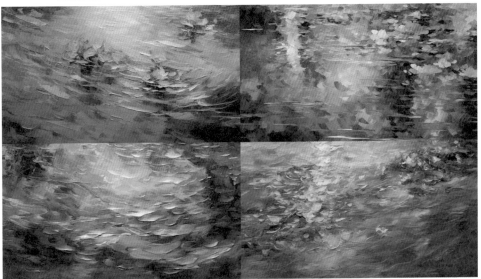

prompt: Reflection effect, shimmering water surface, distorted reflections, vibrant colors, abstract shapes, dreamlike atmosphere, impressionist style, oil paint texture, brush strokes visible, asymmetrical composition, by Claude Monet --ar 16:9 --v 5.1

## 柔和的照明　Soft illumination

prompt: Soft illumination, warm and cozy, autumnal colors, rustic decor, comfortable armchair, crackling fireplace, bookshelf filled with books, steaming cup of tea or coffee, in a vintage film photography style --ar 16:9 --v 5.1

## 柔光　Soft lights

prompt: Soft lights, warm and cozy atmosphere, pastel colors, fluffy blankets, comfortable pillows, flickering candles, rustic wooden furniture, vintage decor, delicate flowers in vases, in a cottage-style interior design --ar 16:9 --v 5.1

## 投影　Projection effect

prompt: Projection effect, immersive and interactive, transforming ordinary objects into extraordinary works of art, dynamic animations and visual effects, creating a multisensory experience for the audience, in a vibrant and colorful style --ar 16:9 --v 5.1

## 發光　Glow effect

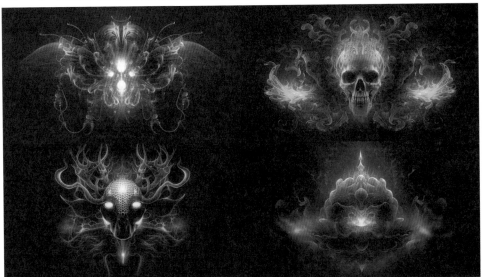

prompt: Glow effect, ethereal and otherworldly, pulsing with energy, surrounded by mist or fog, iridescent colors, floating in a dark void, intricate details, intricate patterns, mysterious and alluring atmosphere --ar 16:9 --v 5.1

螢光燈　Fluorescent lighting

prompt: Fluorescent lighting, a crowded club filled with neon lights and colorful strobes, people dancing wildly to the beat of the music, captured in a dynamic style with fast-paced editing and vibrant colors --ar 16:9 --v 5.1

 浪漫燭光　Romantic candlelight

prompt: Romantic candlelight, soft and flickering flames, delicate glass candle holders, scattered rose petals, elegant table setting, vintage silverware, crystal glasses, intimate and cozy atmosphere, in impressionist painting style --ar 16:9 --v 5.1

柔和燭光　Soft candlelight

prompt: Soft candlelight, warm and cozy, flickering flames, dancing shadows, vintage candlestick holder, delicate lace tablecloth, antique silverware, crystal wine glasses, romantic atmosphere, in impressionist painting style --ar 16:9 --v 5.1

 好看的燈光　Beautiful lighting

prompt: Beautiful lighting, soft and warm tones, gentle shadows, delicate highlights, dreamy atmosphere, bokeh effect, fairy tale setting, vintage style, shot with a Nikon D850 camera, 85mm lens, shallow depth of field, off-center composition --ar 16:9 --v 5.1

## 電光閃爍　Electric flash

prompt: Electric flash, dark and ominous clouds, bright bolts of electricity illuminating the sky, crackling sound effects, intense energy radiating from the storm, chaotic and unpredictable movements, in a dramatic and epic style --ar 16:9 --v 5.1

##  柔軟的光線　Soft light

prompt: Soft light, cozy indoor scene, warm color palette, soft textures, knitted blankets and pillows, flickering candles, steaming hot tea or cocoa, bookshelf in the background, comfortable armchair or sofa, intimate atmosphere, vintage filter --ar 16:9 --v 5.1

## 曖昧光暈　Sultry glow

prompt: Sultry glow, moody and dramatic lighting, a couple embracing in a dimly lit alleyway, shadows cast across their faces, steam rising from a nearby grate, a sense of danger and forbidden love, composition is tight to create intimacy and tension, in a film noir style --ar 16:9 --v 5.1

## 自然光　Natural light

prompt: Natural light, soft and warm, streaming through a window, illuminating a cozy reading nook with a comfortable armchair and a stack of books, creating a peaceful and relaxing atmosphere, in watercolor painting style --ar 16:9 --v 5.1

魔法森林　Enchanted forest

prompt: Enchanted forest, shimmering iridescent leaves, towering trees with twisting roots, misty atmosphere, glowing mushrooms and fireflies, babbling brook , in impressionist style with soft brush strokes and vibrant colors --ar 16:9 --v 5.1

 霧氣朦朧　Misty foggy

prompt: Misty foggy city, skyscrapers shrouded in mist, neon lights and street lamps casting an eerie glow, people hurrying along the streets, a sense of anonymity and detachment, in gritty black and white photography style --ar 16:9 --v 5.1

## 夢幻霧氣　Dreamy haze

prompt: Dreamy haze, soft pastel colors, delicate flowers, floating butterflies, gentle breeze, golden sunlight, dreamlike atmosphere, watercolor painting style, square composition --ar 16:9 --v 5.1

## 仙氣繚繞　Ethereal mist

prompt: Ethereal mist, shimmering and iridescent, hovering over a crystal clear lake, reflecting the colors of the sky, creating a dreamy and surreal atmosphere, swans gliding gracefully on the water, in a soft and romantic impressionistic style --ar 16:9 --v 5.1

溫暖光輝　Warm glow

prompt: Warm glow, golden hour sunlight, soft focus, flowers in foreground, distant mountains in background, dreamy atmosphere, pastel color palette, impressionist painting style, by Claude Monet --ar 16:9 --v 5.1

 憂鬱氛圍　Moody atmosphere

prompt: Moody atmosphere, dark and stormy sky, bare tree branches reaching out like fingers, lone figure standing in the distance with their back turned, silhouette illuminated by a faint light, desaturated colors with a pop of deep red --ar 16:9 --v 5.1

## 柔和月光　Soft moonlight

prompt: Soft moonlight, reflecting off the calm ocean waves, a lone sailboat drifting in the distance, stars twinkling above, sand dunes on the shore,  black and white photography style reminiscent of Ansel Adams --ar 16:9 --v 5.1

## 體積光　Volumetric lighting

prompt: Volumetric lighting, cascading rays of sunshine, filtered through a canopy of leaves, creating a dappled effect on the forest floor, illuminating the vibrant colors of the flora and fauna,   in impressionist style with bold brushstrokes and vivid hues --ar 16:9 --v 5.1

逆光　Back light

prompt: Back light, warm tones of sunrise or sunset, misty atmosphere, trees and foliage in silhouette, leading lines to the horizon, ethereal and dreamlike mood, painterly style inspired by Claude Monet --ar 16:9 --v 5.1

 硬光　Hard light

prompt: Hard light, sharp edges, contrasting shadows, minimalist composition, black and white, high contrast, intense mood, jagged lines, dynamic movement, abstract shapes, in a graphic design style --ar 16:9 --v 5.1

## 林布蘭光　Rembrandt light

prompt: Rembrandt light, dramatic shadows, warm and soft tones, portrait of a man, high contrast, intense gaze, chiaroscuro effect, oil painting texture, classic composition, timeless atmosphere, by Caravaggio --ar 16:9 --v 5.1

 ## 輪廓光　Rim light

prompt: Rim light, dramatic and intense, high contrast, dark background, bright light source behind the subject, emphasizing the edges and contours of the subject, creating a sense of depth and dimensionality, black and white, film noir aesthetic --ar 16:9 --v 5.1

**情調光　Mood lighting**

prompt: Mood lighting, soft focus, dreamy and ethereal atmosphere, pastel colors and delicate details, floral arrangements and decorations, vintage lace and embroidery, candles and fairy lights, intimate setting with a couple in love --ar 16:9 --v 5.1

 **晨光　Morning light**

prompt: Morning light, dew drops on leaves, birds chirping, peaceful atmosphere, misty mountains in the distance, warm colors, gentle breeze, fresh scent of nature, dreamy and ethereal style --ar 16:9 --v 5.1

## 太陽光　Sun light

prompt: Sun light, streaming through the leaves of a tree, dappling the ground with golden spots, creating a serene and peaceful atmosphere, soft breeze blowing, in impressionist painting style, with vibrant colors and loose brushstrokes --ar 16:9 --v 5.1

## 黃金時刻光　Golden hour light

prompt: Golden hour light, warm and glowing, casting long shadows, illuminating a field of wildflowers, soft breeze blowing through the grass, peaceful and serene atmosphere, impressionist painting style, vibrant colors, brush strokes visible --ar 16:9 --v 5.1

## 暗黑的　Moody

prompt: Moody, stormy seascape, towering waves crashing against jagged rocks, dark and ominous clouds overhead, a lone sailboat struggling to stay afloat, oil painting style with bold brushstrokes and rich colors --ar 16:9 --v 5.1

## 鮮豔的　Happy

prompt: Happy, blinding light, ethereal and otherworldly, streaming in through stained glass windows, casting colorful shadows on the ground, illuminating a grand cathedral interior, captured in a soft and dreamy style with a touch of surrealism --ar 16:9 --v 5.1

## 暖光　Warm light

prompt: Warm light, illuminating a vast field of sunflowers, tall stalks swaying in the breeze,distant mountains silhouetted against the sky, a lone farmhouse in the distance, captured in a cinematic style reminiscent of Terrence Malick's films --ar 16:9 --v 5.1

 ## 彩色光　Color light

prompt: Color light, glowing in the dark, reflecting off a wet pavement, a mix of geometric and organic shapes, pulsating with energy and movement, captured in a long exposure shot, wide angle perspective, composition emphasizing symmetry and balance, --ar 16:9 --v 5.1

## 賽博龐克光　Cyberpunk light

prompt: Cyberpunk light, towering skyscrapers with holographic ads, glowing neon lights and signs, dark alleyways and hidden corners, high-tech gadgets and weapons, edgy and futuristic fashion,in a stylized anime-inspired art style --ar 16:9 --v 5.1

 ## 反光　Reflective light

prompt: Reflective light, shimmering and dancing on the water's surface, creating a mesmerizing pattern, soft and dreamy atmosphere, impressionistic style, inspired by Monet's Water Lilies series --ar 16:9 --v 5.1

## 映射光　Mapping light

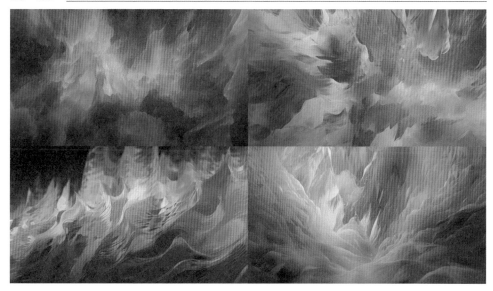

prompt: Mapping light, neon colors, abstract shapes, pulsating rhythms, fluid movements, ethereal atmosphere, otherworldly dimension, holographic textures, glitch effects, in a surreal and dreamlike style --ar 16:9 --v 5.1

## 氣氛照明　Atmospheric lighting

prompt: Atmospheric lighting, soft and diffused, hints of orange and pink, shadows playing across the subject's face, creating depth and mystery, a lone figure standing in a dimly lit alleyway, composition emphasizing negative space --ar 16:9 --v 5.1

殘酷的 Brutal

prompt: Brutal nature, raging storm with lightning strikes and thunderous booms, trees bending and breaking in the wind, waves crashing against rocky cliffs, dark clouds looming overhead, in a painterly style with bold brushstrokes and vivid colors --ar 16:9 --v 5.1

 強烈對比的 Dramatic contrast

prompt: Dramatic contrast, black and white, high key lighting, deep shadows, minimalist composition,film noir style, capturing the subject's face in close-up detail, emphasizing the contrast between light and dark --ar 16:9 --v 5.1

## 陰影效果　Shadow effect

prompt: Shadow effect, mysterious and ethereal, dark and light contrast, soft edges, elongated shapes, monochromatic palette, minimalist composition, in a dreamlike watercolor style --ar 16:9 --v 5.1

## 微光　Rays of shimmering light

prompt: Rays of shimmering light, dancing and weaving through a misty forest, illuminating the trees and foliage with a soft glow, casting shadows and creating a magical atmosphere, captured in watercolor style with pastel colors and gentle brushstrokes --ar 16:9 --v 5.1

強光　Hard lighting

prompt: Hard lighting, neon colors, futuristic setting, abstract shapes and patterns, metallic textures, minimalistic composition,  sharp focus on a single object in the foreground, out-of-focus background, glossy finish, digital art style --ar 16:9 --v 5.1

 冷光　Cold light

prompt: Cold light, icy blue tones, sharp and angular shapes, minimalistic composition, stark contrast between light and shadow, metallic textures, futuristic atmosphere, in a digital art style reminiscent of Tron Legacy --ar 16:9 --v 5.1

## 明亮的　Bright

prompt: Bright summer day at the beach, with crystal clear water reflecting the bright sun overhead.and there are colorful beach umbrellas and towels scattered throughout the scene. A group of friends is playing beach volleyball, The style is realistic and bright, with vivid colors --ar 16:9 --v 5.1

## 雲隙光　Crepuscular ray

prompt: Crepuscular ray, serene beach scene, palm trees swaying in the breeze, golden sand stretching into the distance, waves crashing onto shore, warm and inviting atmosphere, vintage film photography style --ar 16:9 --v 5.1

外太空觀　Outer space view

prompt: Outer space view, vast and endless, stars twinkling in the distance, colorful nebulas and galaxies, a massive planet in the foreground, its rings reflecting light, a spaceship passing by, leaving a trail of light behind it, in a realistic yet surreal style, using digital painting techniques, with an otherworldly atmosphere --ar 16:9 --v 5.1

 分割布光　Split lighting

prompt: Split lighting, dramatic contrast, warm tones on one side and cool tones on the other, subject in the center, sharp shadows, soft highlights, minimalist composition, moody atmosphere, vintage film style --ar 16:9 --v 5.1

前燈 Front lighting

prompt: Front lighting, vintage objects on a wooden table, soft shadows and highlights, warm and cozy atmosphere, shallow depth of field focusing on the foreground, bokeh in the background, captured with a film camera, Pentax K1000, 50mm lens, Kodak Portra 400 film --ar 16:9 --v 5.1

 背光照明 Back lighting

prompt: Back lighting,still life of a vase with flowers on a windowsill, sun rays shining through the window, creating shadows and highlights on the objects, warm and cozy feeling, simple and elegant composition, in oil painting style --ar 16:9 --v 5.1

## 側光　Raking light

prompt: Raking light, dramatic shadows, warm color temperature, wooden texture, vintage objects, antique camera, shallow depth of field, off-center composition, nostalgic atmosphere, in a painterly style --ar 16:9 --v 5.1

## 邊緣光　Edge light

prompt: Edge light, soft pastel colors, dreamy atmosphere, floating objects, surreal landscape, fantasy creatures, glowing orbs, ethereal music playing in the background, creating a sense of wonder and magic --ar 16:9 --v 5.1

## 頂光　Top light

prompt: Top light, soft and diffused, casting gentle shadows, illuminating a cozy living room, warm color palette, comfortable furniture, a bookshelf filled with books, a fireplace crackling in the background,in a rustic and inviting style --ar 16:9 --v 5.1

## 乾淨的背景趨勢　Clean background trending

prompt: Clean background trending,minimalist chic, all-white outfit, sleek and tailored, geometric shapes, clean lines, natural lighting, monochromatic color scheme, modern and sophisticated vibe, overall mood serene and elegant --ar 16:9 --v 5.1

## 邊緣燈　Rim lights

prompt: Rim lights, neon glow, dark background, low angle shot, close-up on subject's face, intense gaze, dramatic shadows, contrasting colors, futuristic vibe, digital art style --ar 16:9 --v 5.1

## 全域照明　Global illuminations

prompt: Global illuminations,ethereal forest, misty and serene, towering trees, dappled sunlight filtering through the leaves, vibrant flora and fauna, a babbling brook running through the center, watercolor painting style --ar 16:9 --v 5.1

## 霓虹燈冷光　Neon cold lighting

prompt: Neon cold lighting,abandoned warehouse, flickering neon lights casting eerie shadows, rusted metal beams and pipes, peeling paint on the walls, broken windows, graffiti-covered surfaces, mysterious atmosphere, in a horror genre style --ar 16:9 --v 5.1

 ## 明暗分明　Stark shadows

prompt: Stark shadows, monochromatic color scheme, minimalist composition, sharp lines and angles, high contrast, eerie atmosphere, black and white photography, vintage film grain texture, by Ansel Adams --ar 16:9 --v 5.1

## 黑暗氛圍 Moody darkness

prompt: Moody darkness, misty forest, towering trees, shadowy figures lurking in the background, moonlight filtering through the canopy, monochromatic color scheme with hints of blue and green, painted in a realistic oil style --ar 16:9 --v 5.1

## 鮮豔色彩 Vibrant color

prompt: Vibrant color, tropical paradise, lush greenery, exotic flowers, crystal clear water, white sandy beaches, colorful birds and wildlife, aerial view, drone photography style, dreamy and surreal atmosphere, in a wide landscape format --ar 16:9 --v 5.1

## 高對比度　Harsh contrast

prompt: Harsh contrast, vibrant colors, neon lights, bold typography, urban landscape, dynamic angles and perspectives, futuristic elements, crowded streets, in digital painting style --ar 16:9 --v 5.1

 ## 安靜恬淡　Serene calm

prompt: Serene calm, misty morning, soft pastel colors, lone boat floating in the distance, gentle ripples on the water's surface, tall trees surrounding the lake, impressionist style with visible brushstrokes, oil on canvas, dreamy atmosphere --ar 16:9 --v 5.1

明亮高光　Bright highlights

prompt: Bright highlights, illuminating a bustling city street at night, casting dramatic shadows and reflections on the wet pavement, surrounded by towering skyscrapers and neon lights, captured in a gritty black and white photography style --ar 16:9 --v 5.1

 閃耀星空　Twinkling stars

prompt: Twinkling stars, vast and endless, shooting stars streaking across the sky, constellations forming patterns, glowing moon casting shadows, peaceful and serene atmosphere, painted with watercolors, dreamy and ethereal style --ar 16:9 --v 5.1

# 場景呈現

第 **6** 章

在茂密的森林中聽樹葉沙沙的聲音，感受生機勃勃的大自然；進入古老的城堡，撫摸粗糙的石牆，領略歷史的厚重與滄桑；登上高樓大廈，眺望繁華的都市，讚歎科技的進步……不同的場景能夠表現出不同的畫面，帶給我們不同的體驗。

## 反烏托邦 Dystopia/Anti-utopia

prompt: Dystopia, abandoned city, overgrown with vegetation, sky filled with smog and pollution, rusty and decaying buildings, oppressive government propaganda posters on every corner, dark and gloomy atmosphere, cyberpunk style --ar 16:9 --v 5.1

## 幻想 Fantasy

prompt: Fantasy,mythical creature, majestic dragon with shimmering scales, soaring high above the clouds, fierce yet elegant, in a vast and beautiful landscape filled with mountains and waterfalls, oil painting style with a touch of realism --ar 16:9 --v 5.1

## 異想天開　Whimsically

prompt: Whimsically tea party, mismatched vintage china, pastel colors, quirky desserts, eccentric guests, floral arrangements, Alice in Wonderland theme, illustrated style, by John Tenniel --ar 16:9 --v 5.1

 ## 廢墟　Ruins

prompt: ancient Ruins, crumbling stone walls, overgrown with vines and moss, scattered rubble and debris,  shafts of sunlight filtering through the cracks, a lone figure standing in the center, gazing up at the sky with a sense of wonder and melancholy --ar 16:9 --v 5.1

## 教室 Classroom

prompt: empty Classroom, muted colors, subtle shadows, minimalistic layout, symmetrical composition, wide angle lens --ar 16:9 --v 5.1

## 臥室 Bedroom

prompt: empty Bedroom, soft morning light, warm and cozy, minimalistic furniture, simple lines, neutral colors, plants in the background --ar 16:9 --v 5.1

 森林　Forest

prompt: A Forest, with no people, in a misty atmosphere, in the style of a fantasy novel illustration, detailed trees and leaves, soft lighting and shadows, --ar 16:9 --c 80 --v 5.1

城市　City

prompt: City skyline at dusk, empty streets, soft colors, warm light, top down view, tilt-shift effect --ar 16:9 --c 30 --v 5.1

## 足球場　Soccer field

prompt: Soccer field, lush green grass, white chalk lines, orange cones, goal posts with nets, players in colorful jerseys, running and kicking the ball,sunny day with blue skies, captured in a dynamic and action-packed style --ar 16:9 --v 5.1

## 體育場　Stadium

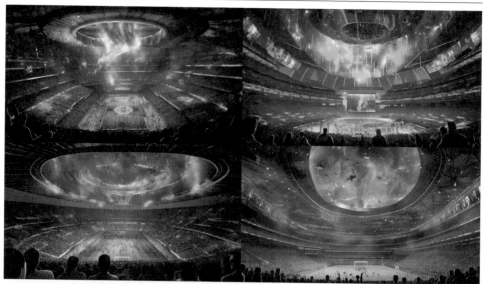

prompt: futuristic Stadium, towering holographic screens, neon lights, sleek and modern architecture, retractable roof, bustling crowd, drones flying overhead capturing the action, intense energy, fast-paced sports game, dynamic camera angles --ar 16:9 --v 5.1

競技場　Arena

prompt: ancient gladiator Arena, sand-covered floor, roaring crowd, fierce warriors, swords and shields, blood and sweat, dramatic lighting, epic battle scene, Roman-inspired architecture, in a historical drama style --ar 16:9 --v 5.1

 摔跤場　Wrestling ring

prompt: Wrestling ring, neon lights, smoke machine, muscular wrestlers, monochromatic color scheme with occasional pops of vibrant colors, in a gritty and raw style reminiscent of 80s wrestling events --ar 16:9 --v 5.1

**泳池　Swimming pool**

prompt: serene Swimming pool, crystal clear water, surrounded by lush greenery, tropical plants, colorful flowers, natural stone tiles, sun loungers and umbrellas, soft and warm lighting, in a watercolor painting style --ar 16:9 --v 5.1

 **咖啡廳　Cafe**

prompt: cozy Cafe, warm lighting, rustic wooden furniture, vintage decor, aroma of freshly brewed coffee, friendly baristas, bookshelves filled with novels and poetry books, in a bohemian style with earthy tones --ar 16:9 --v 5.1

## 麵包店　Bakery

prompt: cozy Bakery, warm colors, rustic wooden tables and chairs, freshly baked bread and pastries on display, vintage decor, hanging plants, natural light pouring in from large windows,in a watercolor painting style --ar 16:9 --v 5.1

 ## 書店　Bookstore

prompt: cozy Bookstore, wooden shelves filled with books, warm lighting, comfortable armchairs, coffee table, fireplace in the corner, aroma of fresh coffee and baked goods, vintage decor, in a watercolor painting style --ar 16:9 --v 5.1

## 居酒屋　Izakaya

prompt: cozy Izakaya, wooden interior, warm lighting, traditional Japanese lanterns, bustling atmosphere, sake bottles lining the walls, sizzling yakitori on the grill,in a rustic and inviting style --ar 16:9 --v 5.1

## 宴會　Banquet

prompt: elegant Banquet, luxurious gold and white decor, chandeliers, floral centerpieces, fine china and crystal glasses,black tie dress code, champagne flowing, delectable gourmet food, captured in a timeless oil painting style --ar 16:9 --v 5.1

音樂會 Concert

prompt: classical Concert, grand piano on stage, orchestra playing in the background, conductor leading the performance with passion and precision, warm lighting casting a soft glow over everything --ar 16:9 --v 5.1

 植物園 Botanical garden

prompt: lush Botanical garden, vibrant colors, exotic plants, winding paths, hidden nooks, trickling water features, romantic atmosphere, impressionist style, soft brushstrokes, dreamy haze effect, oil on canvas --ar 16:9 --v 5.1

## 遊樂園　Amusement park

prompt: colorful Amusement park, whimsical and towering roller coasters, spinning teacup ride, candy-colored cotton candy stand, lively and bustling crowds, bright sunshine, cartoonish and playful style, by Disney Imagineering --ar 16:9 --v 5.1

 ## 教堂　Church

prompt: wedding Church, quaint and charming, white wooden exterior, stained glass windows depicting biblical scenes, lush greenery surrounding the building, romantic atmosphere, vintage film style photography, by Helmut Newton --ar 16:9 --v 5.1

## 中式閣樓　Chinese style loft

prompt: Chinese style loft, minimalistic furniture, natural wood textures, hanging lanterns, calligraphy art on the walls, zen garden with a rock fountain, bonsai trees,in watercolor painting style --ar 16:9 --v 5.1

 ## 花海　Flower ocean

prompt: Flower ocean, vibrant colors, soft petals, gentle breeze, bees buzzing, butterflies fluttering, golden sunlight, flowing water stream, birds chirping, peaceful atmosphere, impressionist painting style --ar 16:9 --v 5.1

## 廢棄城市建築群　Deserted city buildings

prompt: Deserted city buildings, overgrown with vines and foliage, cracked pavement, rusted metal structures, broken windows, eerie silence, post-apocalyptic atmosphere, muted colors, dust and debris scattered around, lone figure standing in the distance, looking towards the horizon, in a surrealistic style --ar 16:9 --v 5.1

## 近未來都市　Near future city

prompt: Near future city, green spaces, clean energy, sustainable architecture, happy and healthy citizens, no pollution, no crime, no poverty, advanced transportation systems, vibrant and colorful atmosphere, optimistic and bright style --ar 16:9 --v 5.1

## 街景　Street scenery

prompt: Street scenery, colorful storefronts, neon signs, crowded sidewalks, busy traffic, towering skyscrapers in the background, a street performer playing music on the corner, captured in a vintage film style with warm tones and grainy texture --ar 16:9 --v 5.1

## 煉金室　Alchemy laboratory

prompt: futuristic Alchemy laboratory, sleek and minimalist design, high-tech equipment and machinery, glowing vials of neon liquid, holographic displays projecting complex formulas and diagrams,in a cyberpunk art style --ar 16:9 --v 5.1

## 宇宙　Universe/Cosmos

prompt: Universe, swirling galaxies, colorful nebulas, sparkling stars, black holes and supernovas, infinite expanse, mysterious and awe-inspiring, in a surreal and dreamlike style --ar 16:9 --v 5.1

## 雨天　Rain

prompt: heavy Rain, urban street at night, neon lights reflecting on the wet pavement, people hurrying under umbrellas, car headlights creating streaks of light, in a painterly style with bold brushstrokes and vibrant colors --ar 16:9 --v 5.1

## 在晨霧中　In the morning mist

prompt: In the morning mist, city skyline emerging from the fog, skyscrapers looming in the distance, a lone figure standing on a rooftop, contemplating the view, monochromatic color scheme with a hint of blue, high contrast and gritty style reminiscent of film noir --ar 16:9 --v 5.1

 ## 充滿陽光　Full of sunlight

prompt: Full of sunlight, crystal clear turquoise water lapping at the shore, palm trees swaying in the gentle breeze, colorful umbrellas dotting the sand, people playing beach volleyball and enjoying the warm weather, dynamic composition with diagonal lines --ar 16:9 --v 5.1

## 銀河　Galaxy

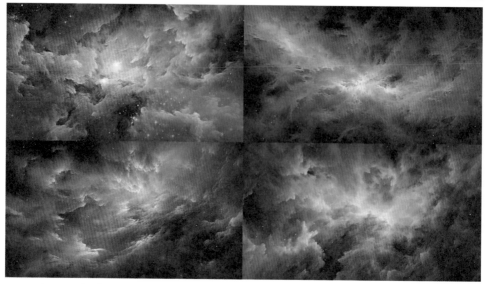

prompt: cosmic Galaxy, swirling colors of purple, blue, and pink, stars twinkling in the distance, vast and endless expanse, glowing nebulae and gas clouds, black hole at the center, in a painterly style with bold brushstrokes and vibrant hues --ar 16:9 --v 5.1

## 地牢　Dungeon

prompt: dark and eerie Dungeon, damp stone walls, flickering torches, rusted iron bars, ominous shadows, cobwebs in the corners, rats scurrying about, musty smell, mysterious runes etched on the floor, in a gothic horror style --ar 16:9 --v 5.1

星雲 Nebula

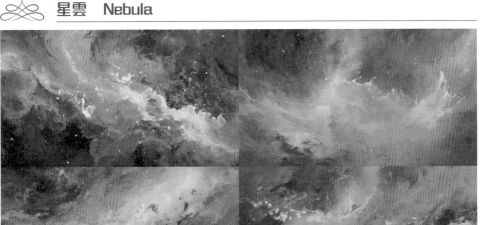

prompt: mesmerizing Nebula, swirling colors of deep purple, electric blue, and fiery orange, interstellar clouds and dust, glowing stars in the background, in a cosmic abstract art style, using acrylic paint on canvas, with bold brushstrokes and splatters --ar 16:9 --v 5.1

 瘋狂麥斯風格 Mad max

prompt: Mad max, vast desert wasteland, scorching sun beating down on the cracked earth, skeletal remains of buildings dotting the horizon, lone figure trudging through the sand with a backpack, muted color palette with hints of warm oranges and yellows --ar 16:9 --v 5.1

## 巴比倫空中花園　Hanging gardens of Babylon

prompt: Hanging gardens of Babylon,lush green oasis, cascading waterfalls, exotic flora and fauna, towering terraces, intricate mosaics, vibrant colors, dreamlike atmosphere, ancient,opulent palace, captured in oil painting style --ar 16:9 --v 5.1

 ## 草原草地　Meadow

prompt: vast Meadow, rolling hills, wildflowers in bloom, clear blue sky with fluffy white clouds, a lone tree in the distance, birds soaring overhead, warm sunlight casting long shadows, peaceful and serene atmosphere, oil painting style --ar 16:9 --v 5.1

**雜草叢生的 Overgrown nature**

prompt: abandoned Overgrown nature, broken down structures covered in vines and moss, rusted metal objects scattered around, rays of sunlight shining through the trees, black and white photography with high contrast and gritty texture --ar 16:9 --v 5.1

 **後啟示錄、末日後 Post apocalyptic**

prompt: Post apocalyptic world, abandoned cityscape, overgrown with vegetation, rusted metal structures, broken glass windows, eerie silence, lone survivor with a gas mask and a backpack, scavenging for resources, ominous red sky, in a gritty and realistic style --ar 16:9 --v 5.1

天空之城　Castle in the sky

prompt: floating Castle in the sky, made of crystal and gold, surrounded by fluffy clouds and rainbows, intricate details and patterns, soaring towers and spires, magical creatures flying around, soft pastel colors, inspired by Art Nouveau style --ar 16:9 --v 5.1

北極光　Aurora borealis

prompt: mesmerizing Aurora borealis, dancing ribbons of green and pink light, swirling and pulsating across a dark sky, snow-covered mountains in the distance, peaceful and serene atmosphere, wide-angle view, expertly composed --ar 16:9 --v 5.1

## 巨大建築　Giant architecture

prompt: Giant architecture, towering skyscrapers, intricate details, gothic style, gargoyles perched on every corner, stained glass windows, ominous clouds in the background, black and white photography, high contrast, dramatic lighting, by Ansel Adams --ar 16:9 --v 5.1

 ## 賽博龐克小巷　Cyberpunk alley

prompt: Cyberpunk alley,neon-lit alleyway, futuristic cityscape in the background, rain pouring down, steam rising from grates, dark and gritty atmosphere, graffiti-covered walls, walking humans, in a Blade Runner-esque style --ar 16:9 --v 5.1

## 星空夜景　Starry night

prompt: Starry night, swirling constellations, vibrant colors, ethereal atmosphere, dreamlike landscape, moonlit sky, silhouetted trees, shooting stars, Van Gogh-inspired style, oil paint texture, impasto brushstrokes, dramatic contrast --ar 16:9 --v 5.1

 ## 數位宇宙　Digital universe

prompt: vast Digital universe, neon colors, pulsating lights, holographic projections, swirling galaxies, intricate networks of data, AI beings, flying through space, futuristic technology, immersive experience, in a cyberpunk style --ar 16:9 --v 5.1

## 超現實夢境　Surreal dreamland

prompt: Surreal dreamland, floating islands of different shapes and sizes, vibrant colors blending together, clouds made of cotton candy, unicorns with rainbow manes and tails, soft and dreamy atmosphere, painted with watercolors in a whimsical style --ar 16:9 --v 5.1

 ## 太空船　Spaceship

prompt: sleek silver Spaceship, aerodynamic curves, glowing blue engine thrusters, rotating radar dish, large cargo bay doors, floating in the vastness of space, surrounded by stars and nebulas, epic sci-fi soundtrack, in a realistic and detailed 3D render style --ar 16:9 --v 5.1

## 懸崖峭壁　Cliff

prompt: Cliff,towering steep cliff, jagged edges, rugged texture, overlooking a vast ocean, crashing waves, misty atmosphere, dramatic lighting, black and white photography style, by Ansel Adams --ar 16:9 --v 5.1

## 神秘森林　Mystical forest

prompt: Mystical forest, misty and ethereal, towering trees with twisted roots, vibrant mushrooms and flowers, glowing fireflies, a hidden path leading to a mysterious clearing, a sense of magic and wonder, in a dreamy watercolor style --ar 16:9 --v 5.1

## 天空島嶼　Sky island

prompt: floating Sky island, lush greenery, crystal clear waterfalls, ancient ruins, mystical creatures, misty atmosphere, pastel color palette, dreamlike composition, inspired by Studio Ghibli's "Castle in the Sky", in soft and painterly style --ar 16:9 --v 5.1

 ## 水晶宮殿　Crystal palace

prompt: Crystal palace, shimmering and translucent, towering spires, intricate and delicate details, surrounded by a moat filled with glittering water, lush gardens with exotic flowers and trees, ethereal atmosphere, soft pastel colors, inspired by Art Nouveau style --ar 16:9 --v 5.1

荒漠孤煙　Desolate desert

prompt: Desolate desert, sand dunes stretching as far as the eye can see, scorching sun beating down, cracked earth beneath feet, sparse vegetation struggling to survive, a lone camel standing tall against the horizon, warm sepia tone, wide-angle lens capturing the vastness of the scene--ar 16:9 --v 5.1

 沉船遺跡　Sunken shipwreck

prompt: Sunken shipwreck, covered in vibrant coral and sea creatures, rays of sunlight shining through the water, rusted metal and broken wood scattered around the ocean floor, schools of fish swimming by, mysterious and eerie feeling --ar 16:9 --v 5.1

## 仙人掌沙漠　Cactus desert

prompt: Cactus desert, towering cacti, golden sand dunes, clear blue sky, warm sunlight, peaceful atmosphere, minimalist composition, fine art photography style, black and white with high contrast --ar 16:9 --v 5.1

## 神話世界　Mythical world

prompt: Mythical world, ethereal creatures with iridescent wings, floating islands with waterfalls cascading down, vibrant flora and fauna, glowing crystals and gems embedded in the landscape, soft pastel colors, dreamlike atmosphere --ar 16:9 --v 5.1

## 外太空　Outer space

prompt: Outer Space,interstellar war, massive battleships, laser beams and plasma cannons, frantic dogfights, explosions and debris, brave pilots in sleek fighter jets, advanced targeting systems and shields, epic space battles  --ar 16:9 --v 5.1

##  魔幻森林　Magical forest

prompt: Magical forest, glowing mushrooms, misty atmosphere, towering trees, hidden creatures, sparkling waterfalls, ethereal music, pastel colors, dreamlike composition, in watercolor style --ar 16:9 --v 5.1

## 古代神廟　Ancient temple

prompt: Ancient temple, overgrown with vines and moss, towering stone pillars, intricate carvings and sculptures, flickering torches casting eerie shadows, mysterious aura, ancient runes etched into the walls, in a dark fantasy style with a touch of horror --ar 16:9 --v 5.1

## 火山噴發　Volcanic eruption

prompt: Volcanic eruption, billowing smoke and ash, fiery lava flowing down the mountainside, ominous red glow in the sky, chaotic and destructive scene, helicopters hovering above for rescue, captured in a dramatic and intense photojournalistic style --ar 16:9 --v 5.1

## 未來機器人　Futuristic robots

prompt: Futuristic robots, with neon lights and glowing eyes, complex circuitry and mechanical parts, hovering above a futuristic cityscape, in a monochromatic color scheme of silver and blue, with a sense of power and sophistication --ar 16:9 --v 5.1

 ## 巨大機器　Giant machines

prompt: Giant machines, towering over the landscape, metallic sheen, intricate gears and pistons, steam billowing out from vents, sparks flying, workers in hard hats and overalls scurrying around,n a realistic hyper-detailed render style --ar 16:9 --v 5.1

## 末日廢墟　Apocalyptic ruins

prompt: post-Apocalyptic ruins, overgrown with vines and moss, rusted metal structures, shattered glass, rubble and debris scattered around, ominous dark clouds in the sky,a lone survivor with tattered clothes and a backpack, in a gritty and desolate style --ar 16:9 --v 5.1

## 星際大戰　Star wars

prompt: Star wars, X-wing fighters and TIE fighters engaged in dogfight, explosions and laser beams filling the screen, Darth Vader's Star Destroyer looming in the background,intense and fast-paced action, futuristic technology and weapons, in a cinematic style --ar 16:9 --v 5.1

## 火星探險　Mars exploration

prompt: Mars exploration, dusty red surface, rocky terrain, rover vehicle,panoramic view of the Martian landscape, astronaut in a space suit collecting samples, orange tinted sky, the sun in the distance, high-tech futuristic design style --ar 16:9 --v 5.1

 ## 科技城市　Technological city

prompt: futuristic Technological city, towering skyscrapers with holographic displays, sleek and aerodynamic vehicles flying through the air, neon lights illuminating the streets, in a cyberpunk style with a gritty and dark atmosphere --ar 16:9 --v 5.1

## 浪漫小鎮　Romantic town

prompt: charming Romantic town, cobblestone streets, colorful buildings with flower boxes,winding alleys, a river running through the town, warm and cozy atmosphere, impressionist style painting, soft brushstrokes and vibrant colors --ar 16:9 --v 5.1

## 蒸汽龐克工廠　Steampunk factory

prompt: Steampunk factory, a maze of pipes and valves snaking through the dimly lit interior, a control room with levers and gauges manned by a top-hatted engineer, conveyor belts carrying raw materials to be processed by the machinery,rough and industrial --ar 16:9 --v 5.1

雨天城市　Rainy city

prompt: Rainy city, colorful umbrellas dotting the streets, reflections of neon lights on rain-slicked surfaces, a couple sharing an umbrella walking hand in hand, cobblestone streets adding to the charm of the city, vintage film camera aesthetic with grainy texture and muted colors --ar 16:9 --v 5.1

 蘑菇森林　Mushroom forest

prompt: Mushroom forest, colorful and playful mushrooms with quirky shapes and patterns, a rainbow of flowers and plants in full bloom, a babbling brook winding through the forest,bright and cheery color scheme, storybook illustration style --ar 16:9 --v 5.1

## 童話城堡　Fairy tale castle

prompt: Fairy tale castle, candy-colored turrets and walls, gingerbread trim, chocolate fountain in the courtyard, unicorns and fairies frolicking in the gardens,hidden rooms filled with toys and games, painted in a cute and charming style --ar 16:9 --v 5.1

 ## 迷人花園　Enchanted garden

prompt: magical Enchanted garden, filled with vibrant flowers and lush greenery, sparkling fairy lights hanging from the trees, a small pond with lily pads and colorful fish, a hidden gazebo covered in vines, all in a dreamy watercolor style --ar 16:9 --v 5.1

## 後啟示錄世界　Post-apocalyptic world

prompt: Post-apocalyptic world, barren wasteland, ruins of skyscrapers, overgrown vegetation, rusted metal debris, ominous dark clouds, lone survivor with a gas mask and makeshift weapons, scavenging for supplies,gritty and realistic style --ar 16:9 --v 5.1

 ## 魔法王國　Magical kingdom

prompt: enchanted Magical kingdom, floating castles, vibrant colors, mystical creatures, cascading waterfalls, sparkling gems, ethereal mist, towering trees, whimsical architecture, in a dreamlike painterly style --ar 16:9 --v 5.1

## 反烏托邦景觀　Dystopian landscape

prompt: Dystopian landscape, desolate and barren, toxic orange skies, rusted metal structures, overgrown with vines and moss, abandoned vehicles, broken glass and debris scattered everywhere, post-apocalyptic vibe, in a gritty and realistic style --ar 16:9 --v 5.1

## 黑暗森林　Dark forest

prompt: haunting Dark forest, twisted and thorny vines, ominous fog, eerie silence broken only by the rustling of leaves, a mysterious figure lurking in the shadows, a decrepit castle looming in the distance, in a gothic style with dramatic lighting and deep shadows --ar 16:9 --v 5.1

## 月球景觀　Lunar landscape

prompt: Lunar landscape, vast and desolate, towering mountains in the distance, deep craters and valleys, scattered rocks and boulders, eerie blue and purple hues, misty atmosphere, glowing auroras in the sky, in a surreal and dreamlike style --ar 16:9 --v 5.1

## 迷失廢墟　Lost ruins

prompt: ancient Lost ruins, overgrown with vines and moss, crumbling stone pillars and archways, mysterious symbols and hieroglyphics etched into the walls, in a realistic and gritty style reminiscent of Indiana Jones movies --ar 16:9 --v 5.1

冰雪王國　Ice kingdom

prompt: Ice kingdom, crystal clear ice palace, towering icicles, shimmering snowflakes,majestic snow-covered mountains in the background,ice sculptures of mythical creatures, elegant ice queen with a flowing gown made of ice shards, in a surreal and dreamlike style --ar 16:9 --v 5.1

 熱帶天堂　Tropical paradise

prompt: Tropical paradise, misty jungle, exotic flora and fauna, ancient ruins, vibrant sunset skies, secluded beach coves, wooden bungalows on stilts over the water, rustic and earthy color palette with pops of bright hues --ar 16:9 --v 5.1

## 後啟示錄荒野　Post-apocalyptic wasteland

prompt: Post-apocalyptic wasteland, abandoned industrial complex, machinery and equipment left to rust and decay, surrounded by a sea of sand dunes, scorching sun beating down on the desolate landscape,watercolor painting style with muted tones and soft edges --ar 16:9 --v 5.1

## 蒸汽龐克城市　Steampunk cityscape

prompt: Steampunk cityscape, towering clock towers, brass pipes and gears, smog-filled air, steam-powered vehicles, flickering gas lamps, towering smokestacks, massive factories, in a sepia-toned vintage style --ar 16:9 --v 5.1

## 奇幻村莊　Fantasy village

prompt: Fantasy village, floating lanterns, winding paths, colorful buildings, lush greenery, hidden waterfalls, magical creatures roaming around, dreamy and whimsical atmosphere, inspired by Studio Ghibli films, painted in watercolor style --ar 16:9 --v 5.1

 ## 賽博龐克街道　Cyberpunk street

prompt: neon-lit Cyberpunk street, towering holographic billboards, rain-slicked pavement, steam rising from grates,  futuristic vehicles zooming past, augmented reality graffiti,  in a gritty and realistic render style --ar 16:9 --v 5.1

## 神秘寺廟　Mystic temple

prompt: ancient Mystic temple, hidden deep in the jungle, intricate carvings and symbols on the walls, flickering torches casting eerie shadows,ancient artifacts and treasures scattered around, in a dark fantasy style with a touch of horror --ar 16:9 --v 5.1

## 古代遺跡　Ancient ruins

prompt: majestic Ancient ruins, towering stone arches and columns,a vast courtyard filled with rubble and debris, the remains of an ancient civilization long gone, captured in a high-contrast black and white style --ar 16:9 --v 5.1

## 沙漠綠洲　Desert oasis

prompt: serene Desert oasis, palm trees swaying in the gentle breeze, crystal clear water reflecting the bright blue sky, sandy beach with colorful umbrellas and lounge chairs, captured in a vintage film style with soft focus and warm tones --ar 16:9 --v 5.1

 ## 月球殖民地　Lunar colony

prompt: Lunar colony, minimalist design, dome-shaped buildings, solar panels and wind turbines, people in space suits walking around, Earth visible in the distance, peaceful and serene atmosphere, pastel color palette, inspired by Wes Anderson's style --ar 16:9 --v 5.1

蒸汽動力機械　Steam-powered machinery

prompt: antique Steam-powered machinery, intricate brass gears and pipes,worn leather belts and pulleys, heavy iron frame, surrounded by thick clouds of smoke and dust, in a gritty industrial setting, sepia-toned vintage photograph style --ar 16:9 --v 5.1

 廢棄太空船　Abandoned spaceship

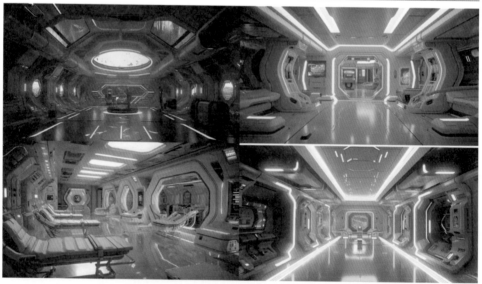

prompt: Abandoned spaceship, sleek and modern design, pristine white walls and floors, advanced technology equipment and machinery, holographic displays and controls, zero gravity environment, neon lighting style --ar 16:9 --v 5.1

## 神秘古墓　Mysterious tomb

prompt: Mysterious tomb, dimly lit by flickering torches, intricate carvings on the walls and pillars, dusty and cobweb-filled corners, a large stone sarcophagus in the center, adorned with precious gems and gold,  in realistic hyper detailed render style, moody, dark tones --ar 16:9 --v 5.1

 ## 冥界　Underworld

prompt: dark Underworld, glowing lava pits, twisted roots and vines, ominous mist, towering jagged rocks, creepy creatures lurking in the shadows, ancient ruins, eerie silence broken only by the occasional rumble of thunder, in a gothic horror style --ar 16:9 --v 5.1

## 霓虹城市　Neon city

prompt: Neon city, towering skyscrapers, flying cars, bustling streets, holographic advertisements, dark and moody atmosphere, cyberpunk style, rain-soaked pavement, reflections on wet surfaces,in a cinematic and dramatic render style --ar 16:9 --v 5.1

## 外星球　Alien planet

prompt: Alien planet, bioluminescent flora and fauna, towering rock formations, misty atmosphere, multiple moons in the sky, strange architecture with glowing symbols, floating islands, in a dreamlike and surreal style --ar 16:9 --v 5.1

## 未來公園　Futuristic park

prompt: Futuristic park, holographic trees and flowers, neon lights, sleek and minimalist benches, geometric sculptures, flying drones, glass pavilions, people wearing augmented reality headsets, in a sci-fi cyberpunk style --ar 16:9 --v 5.1

 ## 時光穿越城市　Time travel cityscape

prompt: Time travel cityscape, neon lights, flying cars, towering skyscrapers, holographic billboards, bustling crowds, time portals, advanced technology, cyberpunk vibe, in a digital art style with vibrant colors and sharp lines --ar 16:9 --v 5.1

## 機器人工廠　Robot factory

prompt: futuristic Robot factory, neon lights, metallic pipes and wires, robotic arms assembling parts, conveyor belts transporting materials, steam rising from machines, sparks flying, in a gritty and industrial style --ar 16:9 --v 5.1

## 賽博龐克叢林　Cyber jungle

prompt: digital Cyber jungle, pixelated flora and fauna, neon lights illuminating the path, futuristic architecture blending with nature,  hidden underground tunnels, in a glitch art style with vibrant colors and distorted shapes --ar 16:9 --v 5.1

## 神秘山脈 Mystical mountain

prompt: Mystical mountain, shrouded in mist and fog, towering peaks, jagged rocks, cascading waterfalls, colorful flora and fauna, tranquil lake at the foot of the mountain, serene atmosphere, ethereal and dreamlike style, painted with soft pastels --ar 16:9 --v 5.1

 ## 冰洞穴 Ice cave

prompt: frozen Ice cave, glittering icicles hanging from the ceiling, crystal clear walls and floors, faint blue light reflecting off the ice, mysterious shadows lurking in the corners, in a surrealistic painterly style --ar 16:9 --v 5.1

## 賽博龐克城市　Cyberpunk city

prompt: Cyberpunk city, sleek and shiny buildings, hovercars zooming through the air, robots and cyborgs walking alongside humans, neon lights illuminating the streets below, a sense of optimism and progress permeating the atmosphere --ar 16:9 --v 5.1

## 反烏托邦未來　Dystopian future

prompt: Dystopian future, desolate wasteland, rusted metal ruins, toxic clouds, mutated creatures, scavengers with cybernetic enhancements, neon lights flickering in the distance, futuristic weapons, gritty and dark atmosphere --ar 16:9 --v 5.1

## 鬼影森林　Haunted forest

prompt: Haunted forest, monochromatic color scheme, stark black and white contrast, twisted trees with gnarled branches reaching out like skeletal fingers, dense fog obscuring the path,in a gothic horror style reminiscent of Edgar Allan Poe's stories --ar 16:9 --v 5.1

## 外星地貌　Extraterrestrial landscape

prompt: Extraterrestrial landscape, barren and desolate terrain, jagged rock formations, ominous dark clouds in the sky, eerie and unsettling atmosphere, harsh lighting casting long shadows, in a black and white photography --ar 16:9 --v 5.1

## 超現實主義風景　Surreal landscape

prompt: Surreal landscape, twisted and distorted trees, giant crystal formations, glowing orbs floating in the air, dark and eerie atmosphere, stormy sky with lightning strikes, abandoned ruins in the background, in a dark fantasy style --ar 16:9 --v 5.1

## 中世紀城堡　Medieval castle

prompt: majestic Medieval castle, towering stone walls, imposing gatehouse, drawbridge over a moat, flags fluttering in the wind,Gothic architecture with intricate carvings and stained glass windows, misty fog adding an air of mystery and intrigue --ar 16:9 --v 5.1

## 月球地貌　Lunar landscape

prompt: Lunar landscape, barren and rocky terrain, towering mountains in the distance, deep craters and valleys, soft glow of moonlight illuminating the scene,futuristic structures dotting the landscape, rendered in a realistic style with high attention to detail --ar 16:9 --v 5.1

## 水晶洞穴　Crystal cave

prompt: mystical Crystal cave, luminescent stalactites and stalagmites, iridescent crystals of all colors, underground river with glowing water, ethereal mist, eerie silence, otherworldly atmosphere,in a dreamy and surreal style --ar 16:9 --v 5.1

## 世界末日城市　Apocalyptic city

prompt: post-Apocalyptic city, overgrown with vegetation, dilapidated buildings, rusted metal structures, abandoned vehicles, dark and moody atmosphere, ominous storm clouds on the horizon, in a gritty and realistic style --ar 16:9 --v 5.1

## 水下世界　Underwater world

prompt: mystical Underwater world, glowing bioluminescent creatures, intricate coral reefs, schools of colorful fish, rays of sunlight filtering through the water,ethereal and dreamlike composition, in a digital painting style with soft pastel colors --ar 16:9 --v 5.1

## 未來都市　Futuristic metropolis

prompt: Futuristic metropolis, towering skyscrapers with holographic advertisements, flying cars zooming through the air, a central plaza with a massive holographic display showcasing the latest technological advancements,in a hyper-realistic render style --ar 16:9 --v 5.1

## 魔法花園　Enchanted garden

prompt: Enchanted garden, glowing flowers, towering trees, winding paths, hidden waterfalls, magical creatures, ethereal mist, soft pastel colors, dreamlike atmosphere, impressionist painting style --ar 16:9 --v 5.1

## 魔法城堡　Magical castle

prompt: Magical castle ,floating in the clouds, made entirely of shimmering crystals, surrounded by colorful rainbows and sparkling stars, guarded by winged unicorns and dragons, in a dreamlike and ethereal style --ar 16:9 --v 5.1

## 夢幻雲彩　Dreamy clouds

prompt: Dreamy clouds, pastel colors, soft and fluffy texture, billowing and swirling shapes, sun rays shining through, birds flying in the distance, peaceful and serene atmosphere, watercolor painting style, inspired by Claude Monet --ar 16:9 --v 5.1

## 工業城市　Industrial cityscape

prompt: Industrial cityscape, towering smokestacks, rusted metal structures, dark skies, polluted air, neon lights, abandoned factories, broken windows, graffiti-covered walls, ominous atmosphere, cyberpunk style --ar 16:9 --v 5.1

 ## 歌德式大教堂　Gothic cathedral

prompt: towering Gothic cathedral, intricate stone carvings, stained glass windows, flickering candlelight,hushed whispers of prayers, misty morning light filtering through the arches, in moody and dramatic style --ar 16:9 --v 5.1

## 浮空城市　Floating city

prompt: ethereal Floating city, made of crystal and glass, reflecting the colors of the sky, surrounded by clouds and mist, intricate architecture with winding staircases and arches, filled with lush greenery and waterfalls, in a dreamy and surreal style --ar 16:9 --v 5.1

## 火星　Red planet

prompt: barren Red planet, desolate landscape with scattered boulders and craters, ominous dark clouds gathering on the horizon, lone astronaut in a bulky spacesuit gazing out at the vast emptiness,  in a gritty black and white photography style --ar 16:9 --v 5.1

## 幽靈莊園　Haunted mansion

prompt: Haunted mansion, dilapidated and overgrown with vines, flickering candles in the windows, eerie mist surrounding the property, ghostly apparitions floating in the air, old-fashioned furniture covered in cobwebs,in a gothic horror style --ar 16:9 --v 5.1

## 海盜島嶼　Pirate island

prompt: Pirate island, misty and mysterious, hidden coves and secret beaches, towering cliffs and rocky terrain, skull and crossbones flag waving in the wind, in a painterly style with bold brushstrokes and vibrant colors --ar 16:9 --v 5.1

未來實驗室　Futuristic laboratory

prompt: Futuristic laboratory, organic shapes and curves, bioluminescent plants and organisms, AI assistants and robots assisting scientists in research and experimentation, warm color palette with touches of gold and copper, in a biomimicry style --ar 16:9 --v 5.1

 太空站　Space station

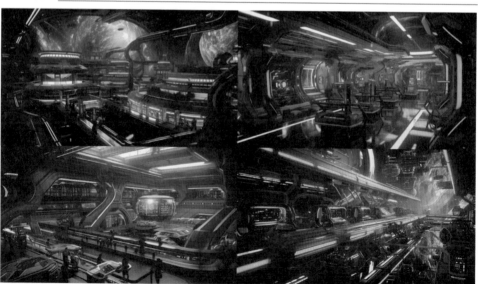

prompt: futuristic Space station, sleek and metallic, with large windows showing the vast expanse of space outside, zero gravity environment, advanced technology and machinery, neon lights and holographic displays, in a sci-fi cyberpunk style --ar 16:9 --v 5.1